The R.A.M.S. Library of Alchemy

Volume 35

Aurifontina Chymica

John Houpreght

R.A.M.S. Publishing Company

Aurifontina Chymica

John Houpreght

14 Alchemical Tracts

Produced by

Restorers of Alchemical Manuscripts Society
1981

R.A.M.S. Publishing Company

R.A.M.S. Publishing Company
117 Rutherford Lane
Stuarts Draft VA 24477

Aurifontina Chymica
Copyright © 2015 R.A.M.S. Publishing Company

First Edition 2015

ISBN-13 **978-1511914260**
ISBN-10 **1511914262**

Image Processing by Philip N. Wheeler

This book is sold for informational purposes only. Neither the publisher nor the editor shall be held accountable for the use or misuse of the information in this book.

Printed in the United States of America

Table of Contents

Dedicated to Hans W. Nintzel,

American Alchemist

and

Founder of the

Restorers of Alchemical Manuscripts Society

(R.A.M.S.)

Disclaimer

Liability: The publisher does not warrant or assume any legal liability or responsibility for the accuracy, completeness, or usefulness of any information, apparatus, product, or process disclosed. The publisher makes no representation as to the accuracy or completeness of the contents of this book and specifically disclaims any implied warranty of merchantability or fitness for a particular purpose. No warranty may be created or extended by written sales materials or sales representatives. You should obtain professional consultation where appropriate. The publisher shall not be liable for any loss of profit or other commercial or personal damages, including but not limited to special, incidental, consequential, or other damages.

Introduction

Philip N. Wheeler

John Frederick Houpreght gathered these fourteen short works and published them in 1680. Original spellings and grammar have been preserved as far as possible.

Notable authors include Raymond Lully, Bernard Trivisan, George Ripley, and Nicolas Flammell. The 14 works are:

Aurifontina Chymica

Hydropyrographum Hermeticum

The Privy Seal of Secrets

A Letter

A Treatise Of Mercury

Colours to be observed in the Operation of the Great Work

Sive Medicina Aurea

Tractatus Delapide

Nicolas Flammell's Summary of Philosophy.

Clavicula by Raymond Lully

Secrets Disclos'd

A Philosophical Riddle

The Answer Of Bernardus Trevisanus

The Prefatory Epistle of Bernard

AURIFONTINA CHYMICA:

OR, A

COLLECTION

Of Fourteen Small

TREATISES Concerning the

First   Matter

OF

Philosophers,

For the discovery of their
(hitherto so much concealed)

MERCURY

Which many have studiously
endeavoured to Hide, but these to make Manifest, for
the benefit of Mankind in general.

LONDON,

Printed for William Cooper, at the Pelican in
Little-Britain, 1680.

HYDROPYROGRAPHUM HERMETICUM

Dear Son, to point out unto thee succinctly a
Memorandum, as it were concerning the understanding
of the true and genuine Stone of the Philosophers,
and the manner of proceeding in its preparation, I
give thee this information, that the said Stone[1] is
compounded and engendered of two things, viz. Body
and Spirit, or of Masculine and Feminine Seed, that
is, of the Water of Mercury, and of the Body of Sol;
whereof we find sufficient proofs and attestations
in all true Writings of the Philosophers, and
therefore I count it needless to enlarge myself by
quoting of them. The upshot of all therefore is,
that first of all Mercury be dissolved and reduced
into a spiritual Water,[2] which is termed by the
Philosophers, the first Matter of Metals, the juice
of Lune, Aqua Vite, Quintessence, a fiery ardent
Water or Brandy; by which Water or prime Matter,[3]
Metals are unlocke'd or untyed, and freed from their
hard and stiff bonds, and reduced into their first
and uniform nature, such as the Water of Mercury
itself is. Upon this account the Philosophers
presented unto us in their Books, the example of
Ice, or frozen Water which by heat is reduced into
Water,[4] because before its coagulation it hath been
Water. Also telling us, that by the very same
principles, from which each thing hath its rise, it
may be reduced or brought back to what it was in the
beginning. And thence they inferr, that it is
impossible to transmute Metals into Gold or Silver,

---

[1] Lapis ex duobus, corpore & spiritus.
[2] Mercurius resolvedus in aquam.
[3] Cum aqua Mercurii metalla resolvenda in primam materiam.
[4] Qualibet res redigitur in primam materiam per principia unde
erta fiat.

without reducing them first into their prima materia.[5] Concerning therefore the bringing about this Regeneration of Metals, thou must diligently heed and observe, my beloved Son, that the same is to be performed only by the means of the prime Matter of Metals, that is, the Water of Mercury,[6] and by nothing besides in the whole World. For this Water is next of Kin unto the Metallick nature, in so much that after their mutual and uniform commixture, they can never be any more parted asunder. This the Philosophers in the Turba and other Books signifie unto us, saying, Nature rejoyceth in its Nature; Nature sustaineth Nature; it amendeth Nature; it reduceth Nature; Nature overcometh Nature. Consequently it is necessary to know this blessed Water, and its preparation, which Water is a hot, fiery,[7] piercing Spirit, the Philosophical Water, and the hidden Key of this Art. For without this all the labour and work of Alchymy is fruitless and frustraneous. Observe therefore, my son, and mark, that all the ground-work of the Philosophers Stone,[8] consisteth in this, that by means of the prima materia metallerum, that is with the Water of Mercury, we reduce and bring back the perfect body of Sol to a new birth, that it be born again by Water and Spirit, according to our Saviour's Doctrine: Except a man be born again of Water and of the Spirit, he cannot See the Kingdom of God. So likewise in this Art, I tell thee my Son, unless the Body of Sol be sowed in its proper soyl, your labour is in vain and it produceth no fruit; as Christ our Saviour saith, Unless a grain of wheat fall into the ground and dye, and rot, it

[5] Redicetio Metallerua in primam materiam guomode fiat.
[6] Aqua Mercurii metallis amica.
[7] Aqua Mercurii est vapor igneus.
[8] Fundamentum Lapidis.

bringeth no fruit. So when the Body of Sol is regenerated by the Water and Spirit,[9] there groweth and cometh forth a clarified, astral, eternal, immortal Body, bringing forth much fruit, and able to multiply itself like unto Vegetables. And to this purpose the Philosopher Roger Bacon speaketh, I do assure you, that if the Astrum do cast and impress its inclination into such a clarified Body of Gold, that it will not lose its power and virtue to the very last assay or judgment: For the Body is perfect, and agreeing to all Elements. But if it be not regenerated, no new, nor greater, nor purer, nor higher, nor better thing can come of it. He that doth not know nor understand this Regeneration of Metals, wroght in nature by the Water and Spirit of the prima materia, ought not to meddle at all with this Art:[10] For in truth, without this, all is but falsities, lyes, unprofitable and to no purpose; yea it is impossible to effect it otherways. Hence is that excellent saying of the Philosophers, That everything bringeth forth its like, and what a man soweth, the same he shall also reap, and no other. And to the same purpose the Philosopher Richardus Anglicus saith, Sow Gold and Silver, that by the means of Nature they may bring Fruit.

Consequently, my Son, thou oughtest to choose no other Body[11] for thy Work but Gold, because that all other Bodies are rank and imperfect. And therefore also the Philosophers made the choice of Gold before all other Bodies, because it is of all things in the World the most perfect, illuminating all other Bodies, and infusing life into them; and because it

---

[9] Corpus per aquam Mercurii sit astrale.
[10] Regenerationem metallorum ignorantes abstineant a Chemia.
[11] Corpus Soiil eligendum ad Chemiam.

is of a fixated incombustible nature, of a constant or abiding root, and Fire proof; also, because (as Roger Bacon saith) the corporal Gold, as to its nobility and perfection, cannot be changed, and is the utmost bound and term of all Natural generation, and there is no perfecter thing in the whole World. The like teacheth also the Philosopher Isaacus Hollandus, saying, Our Stone cannot be extracted from any other but a perfect Body, yea the most perfect in the World. And if it were not a perfect Body, what Stone could be extracted thence?[12] in regard that it must have power to quicken all dead Bodies, to purifie the unclean, to mollifie those that are hard, and harden those that are soft: And in truth, it would be impossible to extract so powerful a Stone out of an imperfect and crasie Body, for a good perfect thing is not to be got from that which is imperfect and unclean: and although many do fancy, that such an extraction may be brought to pass, yet they erre grosly, and are very unwise. Therefore, my Son, observe, that the red Philosophical Sulphur is in the Gold,[13] as Richardus testifieth, and King Calid saith: Our Sulphur is no common Sulphur, but is of a Mercurial nature, fixated and not flying from the Fire. The same all other Philosophers also do witness, that their red Sulphur is Gold.

It is true, my Son, that the Philosophers do say in their Books that the common Gold or Silver is none of their Gold or Silver, in regard that their Gold and Silver is quick or living,[14] but the common are dead and therefore not capable to bring imperfect

---

[12] A corpore perfecto res perfecta extrabitur.
[13] Sulphur tubeum est in auro.
[14] Aurum vulgi non est aurum Philosopherum, & quomodo hoc intelligendum.

Bodies to perfection, nor to communicate unto them the least of their perfection. For if they should bestow some of their perfection upon others, they themselves would be then imperfect, in regard they have no more perfection, than what is needful for themselves. These words of the Philosophers, my Son, are true, and spoken upon very good ground; for it is impossible for common Gold and Silver, to perfect other Bodies that are imperfect, unless as before taught, that the Body of Sol and Lune be born anew, or regenerated by the Water and Spirit of the prima materia, and thereby a glorious, spiritual, clarified, eternal, fixated, subtle, penetrant Body do grow forth, which afterwards hath power to perfect other Bodies which are imperfect. And therefore the Philosophers also said presently after, that those labours are to this end undertaken about their Stone, that its tincture may be advanced and exalted;[15] for it is requisite, that the Stone be digested and carried on to a far greater degree of subtlety and excellency, than the common Gold and Silver possesseth. To this purpose the Philosopher Bernhard speaketh in his Book, in the words following: Though we take this Body just as Nature hath produced it; for all that it is necessary, that by Art, which in this point must imitate Nature, the same be highly exalted in its perfection, to the end that by the means of that superlative accomplishment, and its super-abundant rays, it may be able to perfect and compleat the imperfect Bodies, as to weight, colour, substance, yea as to their Mineral root and principles. But if it should have remained in that degree, wherein Nature left it, viz. in its simple perfection, and not rendered more perfect or exalted, what should the time of

---

[15] Auri tinctura multiplicatur per aeuam Mercurii.

nine months and a half we spend about it serve for?
Arnoldus in his Epistle speaketh home, saying, Gold
and Silver is in our Stone potentially and
virtually, after a powerful, invisible and natural
way;[16] for if it were not so, no Gold nor Silver
could come of it: but the Gold and Silver existing
in our Stone, is better than the common, because it
is living, but the common is dead. and for this
very reason the Philosophers called it their Gold
and their Silver, because it is powerful in their
Stone, active in its essence but not visible common
Gold and Silver; which is also confirmed by Euclides
in the great Rosary, saying, Nothing cometh of a
perfect thing, in regard it is already perfect and
compleat, being so made by Nature. Whereof we have
an example in Bread; which being fermented and
baked, is perfect in its degree or being, having
attained to its intended end, so that it can be
brought to no further fermentation, to make other
Bread of it. The case is the same with the Gold,
which through length of time hath been deduced by
Nature to a fixated and perfect condition:[17] and so
consequently it is impossible by the means of the
simple Gold to perfect other Bodies, unless the
perfect Body be first dissolved and reduced to its
first Matter: which done, it is introverted by our
labour and Art, and reduced into a true ferment and
tincture. Moreover the philosophers do say, that
there is no coming to a good end, until Gold and
Silver be joyned together in one Body. Here, my
Son, thou must understand Lune metaphorically,[18] and
not according to the letter, because the
Philosophers say in their Writings, that Lune is of

---

[16] Aurum Philosophorum potentiale & nirtuale.
[17] Ex corpore Solis perfecto nihil sit, nisi reducatur ad
primam materiam.
[18] Luna metaphorica.

a cold and moist nature, which description they attribute also unto Mercury: and therefore by Lune is understood Mercury, or the prime Matter, which is the Philosophers Lune, or juice of Lune, as is made plain by the excellent and deeply-fathoming piece, the Clangor buccina.

And thus, my Son, thou art instructed briefly, that no profit is to be got by this Art, unless the perfect Bodies by means of the Philosophers Fire, or Water of Mercury, be reduced into their primum Ens, which is a Sulphureous Water, and not Mercury vive, as the Sophisters suppose. For the first matter[19] of Metals is not Mercury vive, but a clammy Sulphureous Vapour, and a viscous Water, wherein the three principles, viz. Salt, Sulphur, and Mercury, are coexistent. Consequently it is necessary to know the true blessed Water of Mercury, or the Heavenly supernatural Fire, whereby the Bodies are dissolved and melted like Ice.   For the knowledge of this is the greatest secret of all,[20] and is wholly and only in the power of God, and is not to be obtained otherwise, but by fervent prayer unto him.

Concerning this the Philosopher Rogerius saith; God hath created Man, and placed him over Nature and all creatures, though himself be natural, and nothing else but Nature, except the Breath which God breathed into him: The very same is to be the judge of the Works, and their nature. This divine Spirit representeth unto the senses and thoughts, in a true Vision as it were, the first principles of Nature, especially since the natural inbred Spirit discovereth some such grounds, whereupon he may

---

[19] Prima materia quid?
[20] Cognitio aqua Mercurii summum est secretum.

surely relye, and in this Work and earnestness of the Spirit, which is of the natural Creation,[21] the divine Adam representeth in us the dissolution of the whole World. And St. Peter by the kindled and burning fiery Spirit of the inbreathed Breath of God, declareth the same very clearly, saying: That the Elements shall melt with fervent heat; the Earth also, and the Works that are therein, shall be burnt up and that there will be a new World, very glorious, excellent and good, as in Apocalypsi is described. And hereupon the Philosopher concludeth, according to what hath been alledged out of St. Peter, that there shall happen a destruction of the Elementary World by Fire. Observe in this Art, that the Fire must perform the like in its type in Nature. Therefore, my Son, set thy thoughts upon this Water, wherewith the Body of Sol[22] (which as Rogerius witnesseth, is a perfect created World) is burnt up, and destroyed and dissolved, that it is not a common Fire, in regard that common is not able to burn or destroy the Gold: but it is a supernatural incombustible Fire, the strongest of all Fires, yea a Hellish Fire, which only hath power to burn the Gold, and to set the same free from its stiff and hard bonds. This supernatural Fire, which hath such a power over the Gold and other Metallick Bodies, is nothing else but the Spiritual, sulphureous fiery Water of Mercury, wherein the Body of Sol is dissolved and burnt up, and of this dissolved and destroyed Body, a new World likewise is created and born, and the Heavenly Jerusalem, that is an eternal, clarified, subtle, penetrant, fixated Body, which is able to penetrate and perfect

---

[21] Vt macrocosmus corrumptur per ignem, ita etiam microcosmus Philosophicus.
[22] Corpus Solis corrumpendum per aquam Mercurii.

all other Bodies. Hence Rogerius saith, As that is
to be a supernatural Fire, which is to break and
melt into one another the Elements of the whole
World; and as out of the broken corruptible Work of
the dissolved Elements, a new Work will be born,
which will be an everlasting Work; even so the Holy
Trinity hath shewed and signified unto us likewise,
a supernatural Fire in the Heavenly Stone. This
supernatural Fire, my Son, the Philosophers have
hidden in their Books in parabolical expressions,
naming the same by innumerable names, and especially
they term it Balneum Mariae, a moist Horse-dung,
Menstruum Urine, Milk, Bloud, Aqua vitae,[23] and the
like.   Fire, saith Bernhardus,[24] make a vaporous
Fire, continual digesting, not violent, subtle,
airy, clear, close, incombustible, penetrant and
vital: and thereupon he speaketh further, Truly, I
have told thee all the manner and circumstances of
the Fire, which only performeth all, and therefore
he bids the Reader, to consider well and often the
words he said concerning the Fire.   Consequently,
he that is wise will easily perceive thence, that
those words are not to be understood of a common,
but of a supernatural Fire; which also Mary the
Prophetess doth hint, saying, that the Element of
Water doth dissolve the Bodies, and make them white.
And concerning this Fire (which he calleth
Menstruum) and its preparation,[25] Raymund Lullie
speaketh in his Testamentum novissimum, in Codice,
in Anima Metallorum, Luce Mercuriorum, Libra
Mercuriorum, de secretis Naturae, de Quinta
Essentia, & in Elucidario Testamenti, c. 4.
saying, that it is not Humane but Angelical to

---

[23] Nomina aqua Mercurii.
[24] Ignis Bernhard.
[25] Loci Lullii de aqua Mercurii.

reveal this Celestial Fire, and that it is the greatest secret of all, how to attain to the knowledge of it. And moreover he saith in figurative expressions, that this Fire is composed of Horse-dung and Calx vive. But what is prefigured by Calx vive, I will expound in another place. And what is signified by Horse-dung, I mentioned before, Viz. that by Horse-dung is meant the Water of the prima materia,[26] for it is warm and moist like Horse-dung; but it is no common Horse-dung, as many ignorant persons do suppose and understand. Hence saith the Philosopher Alanus, the Philosophers called the moist Fire Horse-dung, in which moisture is kept the occult heat, because it is the property of the fire existing in the Horse belly, not to destroy Gold, but by reason of its moisture to increase it. To the like purpose speaketh Alchidonius: Our Medicine must be hidden in moist Horse-dung, which is the Philosophers Fire. And Alanus: Dear Son, be careful in the work of Putrefaction or Destruction, which is to be performed in gentle heat, that is, in moist Horse-dung. Arnoldus de Villa Nova, in the 9th chap. saith, that the heat of Horse-dung is their Fire. So likewise Alphidius: it is digested and buried in the heat of Horse-dung. And Aristotle: the Earth or Body will enjoy no virtue, unless it be sublimed by the means of Horse-dung. And therefore, saith Hermes, roast and cook it in the heat of Horse-dung. And Morienus: if thou do not find in Horse-dung what thou lookest for, thou hadst best to save thy charges. With these agreeth Arnoldus, saying: Let none seek for any other Fire besides this, for it is the Fire of the Wise,[27] the melting Furnace of the

---

[26] Venter equinus Philosophorum est aqua Mercurii.
[27] Aqua Mercurii est Ignis Philosophorum.

Wise, and their Furnace for calcining, subliming, reverberating, dissolving, and performing of Coagulation and Fixation; for this Water dissolveth all Metals, and calcineth them, and melteth itself together with them, both into red and white. In like manner also the Turba and Senior Speaketh: Our Water is a Fire, and our Water is stronger than any Fire, for it reduceth the Body of Gold into a meer Spirit, which the natural Fire is not able to do, though the natural Fire must likewise be had. For then our Water enters into the natural Bodies, and changeth itself into the primogeniel Water, and afterwards into Earth or Powder, which doth more forcibly burn the Gold than the natural Fire; and accordingly Calid saith, It is truly a Fire, which burneth and grindeth all things.

But the manner of preparing this Philosophick Water or Fire, that is, the Aqua Mercurii,[28] the Philosophers have concealed; however Raymond Lullie of all hath written best of it, though in dark expressions. Accordingly, first of all it will be requisite, to purge Mercury from its extraneous humidity and terrene terrestriety, yet so, as not by means of corruptible things; for by such its noble, fruitful, viridescent and generating Nature would be marred. Avicen, Arnoldus, Geber, Raymundus, in Codicillo, and other Philosophers besides, say, that Mercury is best cleansed by subliming it from common Salt, which done, the sublimate to be thrown into warm Water, which will dissolve and sever the Spirit of Salt from it; afterwards the sublimate being dryed and mixed with Salt of Tartar, and forced through a Retort, it will revive again, and this to be done diverse times, and by this proceeding

---

[28] Aqua Mercurii quomodo preparanda.

Mercury will be freed somewhat from its extraneous
moisture and feculency; and Bernhardus towards the
end of his Epistle saith, that this purgation doth
not hurt Mercury, in regard that the hot Water and
Salt do not penetrate into its substance.   But it
is to be noted here, my Son, that in regard that
Mercury is of an uniform indivisible substance,[29] it
cannot be truly and perfectly cleansed by such an
extraneous means, especially because its terrestrial
impurity lies hid in its inmost center, which by no
Sublimation is to be severed thence, as many
ignorant men, though in vain, attempt. And
therefore other means must be used to free Mercury
vive from those bonds, wherewith Nature hath tyed
him uniformly in the bowels of the Earth, and to
reduce him into its primum ens. which is a
Sulphureous spiritual Water, which must be done
without addition of any hetrogeneous thing,[30] as
Rogerius Bacon under the title of Mercury
testifieth, and Raymundus in the Theorica of his
Testament saith, that if it be not putrefied and
opened after the foresaid manner, the Menstruum will
not be worth a Fig. But when the quick Mercury
without any extraneous thing[31] is set free from its
bonds, and dissolved into the primogeneal Water,
then and but then we are capable to cleanse his
inside, and by distillation to sever the Spirit from
the Water, and terrene terrestriety; concerning
which Separation the Philosophers have writ in an
occult stile, such as no conceited person will
easily apprehend, but especially they described it
figuratively in the distillation of Wine. For in
the distillation of Wine appeareth evidently,[32] that

---

[29] Preparatio aqua Mercurii difficilis.
[30] Mercurius in aquam Sulphuream reducondus absque
heterogeneis.
[31] Mercurius Solutus potest purgari.

the Spirit of Wine is mixed with a great deal of
Water, and terrene terrestriety: but by means of an
artificial distillation, the dry Spirit of Wine may
be severed from all the phlegmatick humidity and
terrene terrestriety, in so much that all the Spirit
is severed from the Water existent in Wine, and the
Water from the Earth, and then remain the Lees, out
of which a white Salt is extracted, and joyned again
with the Spirit, and then the Spirit is distilled
and cohobated diverse times, until all the salt be
gone over with it, whereby the Spirit is hugely
fortified and acuated. And in truth, this is a
notable typical discription represented unto us by
the Philosophers, which in the preparation of the
Water of Mercury we ought to imitate: for after its
dissolution we ought in like manner (as hath been
taught of the Wine) by sublimation sever the Water
or Phlegm from the Spirit, and the Spirit from the
Earth, and to rectifie the Earth, and joyn and
distill it together with the Spirit, until all
together come over the Helm. Of which preparation
of this Water, none of all the Philosophers hath
written more clearly nor better than Raymund Lullie,[33]
viz. in Testamento novissimo, as also in the first
Testament, in libro Mercuriorum, libro Q.
Essentiae, & c. where he doth plainly enough
declare, that after the Putrefaction, Separation,
Distillation of Philosophical Spirit of Wine, the
Spiritual Water is to be mixed again and distilled
with its own Earth, that it come over with it; he
declareth also, how this Philosophical Wine or
Menstruum is fortified and acuated with its own
Salt. And further it is to be noted, that this
Water, Menstruum,[34] or Philosophical Spirit of Wine,

---

[32] Exemplum de vini distillatione.
[33] Lullius peritissimus est in preparatione aqua Mercurii.

being thus prepared, doth dissolve or open its own
Body, or Mercury vive, into the primum Ens, or
primogenial Water, whereby it is multiplied without
end, by means of Putrefaction and Distillation. But
what is said of the Extraction of this Water, is
confirmed by Arnoldus de Villa nova, saying: It is
a substance full of Vapours, which containeth in
itself a fat humidity, whereof the Artist severeth
the Philosophers humidity, such as is fit for the
Work, and is as clear as the tears of eyes; wherein
dwelleth the Quintessence in a Metallick Nature,
very proper for the Metals, and therein is the
Tincture to bring forth an entire Metal: for it
containeth the nature both of Argent vive, and also
of Sulphur. Rosarius Philosophorum saith concerning
the distillation of the Menstruum[35] or Water that
great industry and care must be had, and that the
Vessels to be used for the cleansing of this Spirit,
must be of Glass, and exactly closed, to the end
that the Spirit may find no vent nor place to fly
through, it being very forward to make its way
through any hole it findeth: And if the red Spirit
should be gone, the Artist will lose his labour:
the Philosophers call the red Spirit Bloud, and
Menstruum; therefore be very careful to have good
Vessels, and to have the joynts well luted, that you
may get the dry Spirit with its Bloud into the
Receiver by itself, without evaporation of its
Virtue, and keep it, until thou have occasion to
work with it. But concerning this distillation, the
ocular inspection goeth beyond writing, and none can
be a Master, before he have been a Scholar or
Apprentice. Be provident therefore and discrete in
thy Work,[36] lay on a Receiver, and first distill by a

---

[34] Mercurius Solutus solvit suam corpus.
[35] Vasa bone claudenda inpraparatione aqua Mercurii.

gentle Fire the Element of Water, which being over, put it aside, and lay on another Receiver, and close the joynts exactly, that the Spirit may not vapour away, increasing the Fire a little, and there will rise in the Helm a dry yellow Spirit: Continue the same degree of Fire, so long as the Spirit cometh yellow. But when the Alembick beginneth to be red, then increase the Fire very gently, and keep it going on thus, until the red Spirit and Bloud be quite come over, which in its, ascending and going through the Helm will appear in the form of Clouds in the Air. And so soon as the red Spirit is distilled over, the Helm will be white, and then cease quickly; and thus you have in the Receiver the two Elements of Air and Fire, having extracted the true dry Spirit, and severed the pure from the impure. Loe now you have the prima materia Metallorum, wherein the Bodies are reduced. For all the Metals have their rise from Water,[37] which is a root of all Metals. And therefore they are reduced into Water, like as the frozen Ice by heat is reduced into Water, because it hath been Water before.

Do not marvel at it, for all things upon Earth have their root and rise from Water. O how many there are that work and never think upon the root, which is the Key to the whole Work: it dissolveth the Bodies readily;[38] it is Father and Mother; it openeth and shutteth, and reduceth Metals into what they have been in the beginning. It dissolveth the Bodies, and coagulateth itself together with them; the Spirit is carried upon the Water, that is, the Power

---

[36] Processus in distillatione aqua.
[37] Omnia metalla ex aqua.
[38] Effectus aqua.

26

of the Spirit is seen there operating, which is done
when the Body is put into the Water. Whereupon the
Philosopher saith: Look upon that despicable thing,
whereby our Secret is opened. For it is a thing
which all know well, and he that knoweth it not,
will hardly or never find it: the wise man keepeth
it, and the fool throwes it away, and the reduction
is easie to him that knows it. But my Son, it is
the greatest secret to free this Stone,[39] or Mercury
vive, from its natural bonds, wherewith he is tyed
by Nature, that is, to dissolve and reduce it into
its primogenial Water; for without this be done, all
will prove but labour lost: for else we should not
be able to sever and extract the true Spirit or
Watery substance, which dissolveth the Bodies. And
this Solution hath been concealed by all the
Philosophers,[40] who left it unto God Almighty's
disposing, anathematizing that man that should
openly reveal it. And therefore they speake very
subtilly and concisely concerning the solution of
this crude Body, to the end it may remain occult
unto the unwise.[41] But, my Son, thou art to take
notice, that the solution of Mercury vive will
hardly be performed without a means, but none such
are to be used as are Sophistical, as many rude,
unwise and ignorant fools use to do, who by strange
extravagant ways reduce Mercury into Water,
supposing that to be the right Water. They sublime
Mercury with Corrosives,[42] with all sorts of Salts
and Vitriols, from which the sublimed Mercury
attracteth the Salty Spirits, and then afterwards
they dissolve the sublimate into Water in Balneo, or
in the Cellar, or diverse other ways. Item, they

---

[39] Resolutio Mercurii maximum Secretum.
[40] Omnes Philosophi occultarunt confectionem aque.
[41] Resolutio Mercurii non fit absque medico.
[42] Modi falsa resolutionis Mercurii.

reduce it into Water by Salt-Armoniack, by Herbs, Sope, Aquafort, by means of strange kinds of Vessels, and many the like Sophistical proceedings, all of which are but gross fancies, foolish and frustraneous labours: Some also conceive to sever those things afterwards from the Water of Mercury, and that then it shall be the true Water, which the Philosophers do desire. The reason of their Errors is,[43] that they consider not the words of the Philosophers, who plainly do say, that it ought not to be mixed with any heterogeneous thing in the whole World. And Bernhardus saith in his Epistle, that so soon as Mercury is dryed up by the Salts, Aquafort, and other things, that thenceforward it is unfit for the Philosophick work; for being dryed up by the Salts, Allums, Aquaforts, it is not able to dissolve. But, my dear Son, observe what now I tell thee, and what information concerning this point the Philosophers left behind them in their Books;[44] viz. that the Water is not to be prepared by any heterogeneous means whatever in the whole World, but only by Nature, with Nature, and out of Nature. These words are all plain to the understanding, which I will not now openly unfold, but reserve the same for a peculiar Treatise; however for a Memorandum, I will set down these following Rhimes.

Take fresh, pure, quick, white and
    Clear,
Tye him hands and heels so near, With
a most puissant cord and yoke, That he may be
mortified and
    choakt. Reduce him by his like
homogeneous
    Nature, To his first being, or

---

[43] Causa errorum in confectione aqua.
[44] Hac optime notania.

                primogeneous
        feature, Within the close
Chamber or House
        of putrefaction,
According to Dame Nature's indication: Then you
will have a living Spiritual
        Fountain, Flowing bright and clear from
Heaven's
        Mountain,
Feeding on its proper flesh and bloud, Therewithall
increasing to an endless
        Floud.

Let him, that by Divine assistance obtaineth this
blessed Water, render thanks unto God, for he hath
the Key in his hands, wherewith he may open the fast
Locks of all Metallick Chests,[45] out of which Gold,
Silver, Gems, Honour, Power, and Health are to be
had. This blessed Water is by the Philosophers
called, the Daughter of Pluto, having all the
Treasures in her Power. It is also termed the
white, pure, delicate, undefiled Virgin Beja,
without which no generation nor increase can be
effected. And therefore the Philosophers espoused
this delicious pure Virgin unto Gabricius, to the
end they may raise up Fruit; and when Gabricius lay
with her, he dyed, and Beja out of excessive love
swallowed and consumed him, as Arisleus in Turba
Philosophorum speaketh of it. And Bernhard in his
Practica. saith: the Fountain is as a Mother unto
the King, for she doth attract him, and causeth him
to dye, but the King by her means riseth again, and
uniteth himself so firmly unto her, that no man can
hurt him. And therefore the Philosophers say,
although Gabricius be costlier, dearer, and more
esteemed by the World than Beja, yet he alone can

---

[45] Aqua clavis artis.

bring no Fruit. This Virgin and blessed Water the
Philosophers named in their Books with many thousand
names;[46] they call it Heaven, Celestial Water,
Celestial Rain, the dew of Heaven, May-dew, Water of
Paradice, parting Water, Aqua Regis, a corrosive
Aquafort, sharp Vinegar, Brandy, Quintessence of
Wine, growthful green Juice, a growing Mercury, a
viridescent Water, and Leo Viridis, Quick Silver,
Menstruum, Bloud, Urine, Horse-piss, Milk, and
Virgins Milk, white Arsnick, Silver, Lune, and juice
of Lune, a Woman, Feminine Seed, a sulphureous
vapouring Water and smoak, a fiery burning Spirit, a
deadly piercing poyson, and Basilisk that killeth
all, a venomous Worm, a venomous Serpent, a Dragon,
a Scorpion devouring his Children, a hellish fire of
Horse-dung, a sharp Salt, and Salt-Armoniack, a
common Salt, sharp Soap, Lye, a viscous Oyl,
Ostrich's Stomach which doth devour and concoct all,
an Eagle, Vulture, Bird of Hermes, a Vessel and Seal
of Hermes, a melting and calcining Furnace, and
innumerable other names of Beasts, Birds, Herbs,
Waters, Juices, Milk, Blouds, Etc. And they writ
figuratively in their Books of this Water, to be
made of such things, whereas all the unwise which
sought it in such like things, have not found the
true desired Water. Know therefore, my dear Son,
that it is only made of Mercury vive,[47] and no other
heterogeneous thing in the World; and that the
Philosophers therefore gave it so many Names, that
it might not be known to the unwise. And with this
Item I will conclude this Treatise, whereby thou
mayst understand and learn, that without this Fire
all the labour of the whole World is meerly lost,
all Chymical processes false, lying and useless.

---

[46] Varia appellationes aqua Mercurii.
[47] Aqua Mercurii unassiat.

The great Rosary saith, there is no more but one
Receipt, and with this one Lock all the Philosophers
Books both particularly and universally are lock'd
up, and walled about, and fenced as it were with a
strong Wall; and he that knows not the Key, nor hath
it in possession, is not able to open the Lock, nor
to obtain Fruit. For this Water is the only Key for
to open the Metallick Walls and Gardens. And this
Water is the strong Aquafort, of which Isaacus in
his particular Work is to be understood,
wherewithall he dissolveth and spiritualizeth the
Bodies. And therefore it is very diligently to be
noted, that without this Water nothing can be
effected in Chemica,[48] and without it all are but
falsities and lyes, both in Metals and Minerals, as
also in Vegetables and Animals. Whether they
dissolve, sublime, distill, calcine, extract. mix or
compound with any other thing whatsoever; whether
they dissolve per deliquium in Balneo, in Horse-
dung, in Aquafort, and all sorts of strong Liquors,
which seem to promise some probability, and
according as the pretended processes of Alchymists
do teach or may be invented: whether there be made
Oyl, Water, Calx, Powder, black, white, yellow and
red; whether it be burnt, melted, or done anything
about it, which the Alchymists Receipts do teach and
vent for true, whereby to make Gold and Silver, all
proveth but false and a cheat in the event. For
myself with my own hand have experimented all such
things to my damage and loss, not believing them to
be false before I tryed them. Therefore be exhorted,
my Son, to shun such Sophisters,[49] Cheats and
Imposters, as much as the grand Imposter the Devil,
and avoid them as carefully as a terrible burning

---

[48] Absque aqua nihil fit in Chemia.
[49] Sophista in Chemia vitandi.

Fire, and Poyson; for by such Sophistry, and sweetly insinuating false Alchymy, a man runs the hazard of Body and Soul, Reputation and Wealth, yea this Imposture is worse than the Devil himself. For though a man should spend a whole Province or Kingdom upon such deceitful processes, yet all would be consumed in vain, and no firm truth thereby be obtained. Wherefore open thy eyes, own and acknowledge the only Key, and flee from all falsity; for it is impossible else to speed, or do any good.

FINIS.

THE PRIVY SEAL OF SECRETS

WHICH
Upon pain of Dammnation
is not unadvisedly to be
broken up, nor Revealed to
any but with great Care, and
many Cautions.

To omit circumstances, the first Matter out of which
the Philosophers Stone is to be had and taken, is a
subject common and poor in outward appearance, and
therefore it is called a little thing, and it is in
every Mine, yet is nearer in some things than in
others, and in a word in the Mineral Kingdom you
must have it, in the most excellent work of the
Mineral Hierarchy; therefore not Animals or
Vegetals. Know ye then, (although I deny not
Raymonds Cannons to be true) that the lively Nature
being constrained with the strength of Gold, in the
most subtle heat, the Tincture may be made well
easily, and in a short time, which will convert all
metals into perfect Gold; but the way of the
Philosophers in the Universal Work, was out of the
Mineral Kingdom; leaving therefore Animals and
Vegetals, I will acquaint you with the Universal
Subject. Know that all Philosophers affirm, that the
Matter is but one thing, and a vile thing which
costeth nothing, cast in High-ways and trodden upon,
which is the hope of Metals, or a thing containing
all things needful for the Work within itself; and
albeit curious Wits hold all these to be Aenigma's,
yet they are true according to the letter. Briefly,
to manifest the truth, you shall know that in all

Mines whatsoever there doth lye certain Beds, of a lutinous or clayish substance, under the Earth, which in some places is harder than in others, the deeper the Mine is, the more unctuous is the Clay; and this Clay is the Mother of the Metals, the feeder, of the Mines, for in it lies hid the Spirits, or the three Principles of Metals, (viz) Salt the Body, Sulphur the Soul, and Mercury the Spirit, not common nor running, but a white Vapour which resolves itself into a white Water; I say invisibly in this confused lump of Clay, lies hid the aforesaid Principles.

And this is the true Matter or Subject of the Philosophers, and mark how that it agreeth with that I said before: First, that it is one thing, which yet containeth three; Secondly, that it is a vile thing, and yet is not so, for it is a lump of Clay; Thirdly, that it is so vile and common that Workmen throw it out of their Mines, and tread on it, as a thing of no value: I have seen High-ways paved with it in Hungary, and it is no other in other Countries. And is not this Chaos or confused matter? Is not this the hope of Metals? be you judge. I took my Matter in Hungary out of the Mines of Sol, and so I was taught, because more decocted, and riper or hotter Spirits are there, than in any other Mines. Paracelsus out of it wrought his Elixir, but the Philosophers generally took their Matter (which is the same in shew and substance, but not so ripe) out of the Mines of Saturn, and that is their Saturn so often mentioned in their Books; not Ore of Saturn, nor Mercury of Saturn but the Sperm, where the Vegetable Spirits are not specificated to Lead, but lye hid in the lutinous lump of Clay.

Now the difference between that which is taken out of the Mine of Sol, and that which is had out of the Mine of Saturn, is this; in Sol the Matter is so prepared, you shall have need of but one Putrefaction, but in that taken out of the Mine of Saturn, you must have three Putrefactions, which indeed is the great and universal Work. And thus I have fully and plainly revealed the Matter, the Work is easie, viz.

The Practice.

Take this lutinous Clay out of Sol or Saturn, (for the working in either the Preparation is alike) I say, take that which is most clammy or unctuous, and when you gather it, keep it from the Air, as close as you can in a glass or Earthen Vessel, for it will (which I have admired) in an instant indurate and harden: But put it in a Glass Vessel, and in that digest it being well stopped in B.M. or in a Blind Head which is better; but let three parts of the Vessel be empty, and let the heat of your Balneum be such as you may easily hold your hand in it. Some Philosophers digested this a Philosophical month, which is six weeks, but then their Matter was not fresh; for if it be fresh, then fifteen or twenty days is sufficient. After Digestion alter the head, and distill, and you shall have the Philosophers Oyl; which being come, pour it on the Matter again, and this till you have so much Vinegar as will swim four fingers over the Matter; then let it stand twenty-four hours, and it will be tinged yellow; pour that gently off, and distill away your Vinegar till it come to a gummy substance; then pour this Vinegar on the Matter again, and it will be tinged yellow: distill and reiterate this until your

Vinegar be no more tinged yellow, then hath it sucked out all the Spirits out of the Clay: then from the yellow Liquor distill away all the Vinegar, and you shall have a gummy substance like Saccarum Saturn: digest this two days, then distill away all the Flegm in Balneo, then let it cool, and put it in a Retort, with a great Receiver well luted to it as can be; put it into an Ash Furnace, and distill it again, and by degrees you shall have all your Receiver become as white as Milk, which is crude Mercury of Philosophers, or the Virgin Milk: continue Distillation, and a bloud red Oyl shall ascend, which is Sulphur of Philosophers incombustible and unctuous: continue till no more will come over, with so violent a heat for twelve hours, that you do almost melt the Glass; then let it cool, and take off that Receiver, and stop it up very close: break the Retort, and the Feces will be as black as Pitch, and hard, which grind small on a Marble, then Reverberate it in an Earthen Calcining-pan, close covered for three days, (but make not the Matter red-hot) and lay it two fingers thick in the Pan: then take it out, and either with your Vinegar rectified from its Feces, or with Rainwater distilled, I have tried and found it being well Reverberated, that it will take up the Salt, held the Vinegar the best and most proper; digest it therefore with Vinegar twenty hours, then filter and distill it in B.M. till it dry; dissolve it again in that Vinegar, but first rectifie it: let it settle, philter and distill, and reiterate until the Salt be Crystalline and white, then put it in a white glass Body, pour thereon this red Oyl which is the Sulphur, and also the white Water which is incorporated therewith: lute on close and well a Blind Head, and digest in Balneo three days and it

will be all one thing or pap: but then distill away
all the humidity that will arise, and then put it in
an Egg-glass with a short neck, nip it up without
heating the Matter, let the Egg be but a fourth part
full.

This is the gross Conjunction and Preparation,
without adding any more than Natures proportion:
put the Glass in an Athanor, in a gentle heat, and
the Matter shall dissolve, putrifie, and perform all
the Work by vertue of Count Trevisan's Fire, which
is the Spirit ever working within the Glass,
beginning visibly before the Matter begins to
putrifie, for these it continually ascends and
descends until Congelation. Be not too curious,
only pray to God, and he will direct your Work, and
bring it to a period, which I judge to be sixteen
months, a bloud-red Powder impalpable in the
conclusion of the Work, be patient and you cannot
erre. Note, I was never taught to multiply, but by
increasing with his own Oyl and Salt, that is, with
ten parts Oyl and one of Salt depurated, and so
increasing the Medicine you shall bring it as high
as you will. I know not any more than this, neither
can any more large or more plainly. Serve God, and
you cannot erre.

Know also, that you may with this Fire-Stone, which
is the red Oyl, and this Salt prepared from the
white Water, increase Procipitate of Sol and
Mercury, elevated together and then mixed, or upon a
subtle Calx of Sol alone, but not so suddenly. The
mannor is, to pour on the Calx the red Oyl, till it
be like pap; then lute it, and set it in Ashes to
circulate in a Circulatory, that if any Mercurial
Spirit should remain, it may still arise and not

hinder the fixation of the Matter: continue the
Fire till it be a dry Powder, then increase more and
more, till it be in an Oyline substance fixt, which
turns Luna into perfect Sol with great profit. And
thus you may increase with the Oyl of Antimony, as I
have shewed you.

Finis.

A LETTER

Communicated by the most Serene Prince

FREDERICK

Duke of

Halsatia and Sleswick, Concerning an Adept,

AND

Relates things strange and unheard-of.

My Friend,

You have desired of me an account of the Life and
Death, Inheritance and Heirs of my Master B.J. of
happy memory: I return you this Answer, in Latine,
as yours to me was, though I be not exactly skilled
in it.

He was by Nation a Jew, by Religion a Christian, for
he believed in Christ the Saviour, and openly made
profession of the same: He was a man of great
Honesty, and gave great Alms in secret: He lived
chastly a Batchelor, and took me when I was about
twenty years of Age, out of the House where Orphans
are maintained by the Publick, and caused me to be
instructed in the Latine, French, and Italian
Tongues; to which I afterwards by use added the
Jewish or Hebrew. He made use of me, so far as I was
capable, in his Laboratory, for he had great skill
in Physick, and cured most desperate Diseases. When
I was twenty five years of Age, he called me into

his Parlour, and made me swear to him, that I would never marry without his consent and knowledge; which I promised, and have religiously kept.

When I was thirty years of Age, on a morning he sends for me into his Parlour, and said very lovingly to me, My Son, I perceive that the Balsom of my Life, by reason of extreme old Age coming on, (for he was eighty eight years of Age) is well-nigh wasted, and that consequently my Death is at the door: wherefore I have writ my last Will and Testament, for the use and benefit of my Brothers Sons, and of you, and have laid it upon the Table of my Closet, whither neither you nor any mortal ever entered: for you durst not so much as knock at the door, during the hours set apart for my Devotion. Having said this, he went to the double door of his Closet, and daubed over the joynings thereof with a certain transparent and Crystalline Matter, which he wrought with his fingers till it became soft and yielding like Wax, and imprinted his Golden Seal upon it; the said Matter was immediately hardened by the cold Air, so that without defacing the Seal, the door could no way be opened.

Then he took the Keys of the Closet, and shut them up in a small Cabinet, and sealed the same as before with the said Crystalline Matter, and delivered the said Cabinet, after he had sealed it, into my hands, and charged me to deliver the same to none but his Brothers Sons, Mr. Jesse, Abrah, and Solomon Joelha, who at that time lived in Switzerland, the eldest of them being a Batchelor.

After this he returned with me into the Parlour, and in my presence dropped the Golden Seal he had made use of, into a glass of clear Water, in which the said Seal was immediately dissolved, like Ice in hot Water, a white Powder settling to the bottom, and the Liquor was ting'd with pale red of a Provence Rose. Then he closed the said Glass Vial, with the above-mentioned transparent Matter, and charged me to deliver the said Vial, together with the Keys, to Mr. Jesse.

This being done, he repeated upon his bended knees some of Davids Psalms in Hebrew, and betook himself to his Couch, where he was used to sleep after Dinner, and commanded me to bring him a Glass of Malaga, which now and then he sparingly made use of: As soon as he had drank off his Wine, he bid me come to him, and leaning his head upon my shoulders, he fell into a quiet sleep, and after half an hours time fetched a very deep sigh, and so yielded his Soul to God, to my great astonishment.

Upon this I according to my promise writ into Switzerland, to give notice of his death to his Nephews; and to my great wonder, the very day after my blessed Master died, I received a Letter from Mr. Jesse, wherein he enquired whether my Master were dead or alive, as if he had known everything that had passed; as indeed he did, by means of a certain Instrument, of which hereafter I shall make mention.

A little after his Nephews came, to whom I gave an account of what had passed; all which Mr. Jesse heard with a smile, but the other Brother not without astonishment and wonder. I gave him the Keys, together with the Glass in which was the aforesaid Golden Solution; but they refused then to meddle with anything that day being tired with their

Journey, but on the morrow, after I had carefully shut all the doors of the house, and none but they and I being present, Mr. Jesse took the Glass Vial, and broke it over a China-dish, which might receive the inclosed Liquor and put it upon the transparent Matter, with which the Cabinet was sealed, and immediately the Matter which before was hard as chrystal, was resolved into a thickish Water; so he opened the Cabinet, and took thence the Keys of the Closet.

Then we came to the door of the Closet, where Mr. Jesse having seen the Seal, he wetted it as formerly with the forementioned Liquor, which immediately gave way; and so he opened the said double door, but shut it again, and falling down upon his knees, prayed, as we also did; then we entered, and shut the doors upon us. Here I saw great Miracles.

In the midst of the Closet stood a Table, whose Frame was of Ebony: the Table itself was round, and of the same Wood, but covered with Plates of beaten Gold; before the Table was placed a low Footstool, for to kneel upon: in the midst of the Table stood an instrument of a strange and wonderful contrivance, the lower part of it or Pedestal was of pure Gold, the middle part was of most transparent Crystal, in which was inclosed an incombustible and perpetually-shining Fire; the upper part of it was likewise of pure Gold, made in the form of a small Cup, or Vial.

Just above this Instrument hung down a Chain of Gold, to which was fastned an artificial Crystal, of an Oval form, filled with the aforesaid perpetual Fire.

On the right side of the Table we took notice of a Golden Box, and upon the same a little Spoon: this Box contained a Balsom of a Scarlet colour.

On the left side we saw a little Desk of Massie Gold, upon which was laid a Book containing twelve leaves of pure beaten Gold, being tractable and flexible as Paper; in the midst of the leaves were several Characters engraved, as likewise in the Corners of the said leaves, but in the space between the Center and corners of the leaves, were filled with Holy Prayers.

Under the Desk we found the last Will of my deceased Master; whilst we were in the Closet, Mr. Jesse Kneeled down, leaning upon the Desk, and with most humble devotion repeated some of the forementioned Prayers, and then with the little spoon took up a small quantity of the aforesaid Balsom and put it into the top of the Instrument which was in the midst of the Table, and instantly a most grateful Fume ascending which with it most pleasing odour did most sensibly refresh us: but that which to me seem'd miraculous was, that the said Fume ascending, caused the perpetual Fire enclosed in the hanging Crystal, to flash and blaze terribly, like some great star or Lightning.

After this, Mr. Jesse read the Will, wherein he bequeathed to Mr. Jesse all his Instruments and Books of Wisdom, and the rest of his Goods to be equally divided between him and his Brother; besides he left me a Legacy of 6000 Golden Ducatoons, as an acknowledgment of my fidelity.

And accordingly first enquiry was made for the Instruments and Books of Wisdom; of those that were on and about the Table, I have spoken already: in the right side of the Closet stood a Chest of Ebony, whose inside was all covered with Plates of beaten Gold, and contained twelve Characters engraven upon them.

From thence we went to view a large Chest, containing twelve looking-glasses not made of Glass, but of a certain wonderful unknown Matter; the Center of the said Looking-glasses were filled with wonderful Characters, the Brims of them were inclosed in pure Gold, and between the said Brims and center they were polished, looking-glasses receiving all opposite Images.

After this we opened a very large Chest or Case, in which we found a most capacious looking-glass, which Mr. Jesse told us was Solomons Looking-glass, and the Miracle of the whole World; in which the Characterisms of the whole Universe were united.

We saw also in a Box of Ebony, a Globe made of a wonderful Matter; Mr. Jesse told us, that in the said Globe was shut up the Fire and Soul of the World, and that therefore the said Globe of itself performed all its motions, in an exact Harmony and Agreement with those of the Universe.

Upon this Box forementioned stood another, which contained an Instrument[50] resembling a Clock-Dial, but instead of the Figures of the 12 hours, the Letters of the Alphabet were placed around this,

---

[50] See this Instrument described in a Book called Arts Notoria, Printed in Latine or English, page 136.

with a Hand or Index turning and pointing at them.
Mr. Jesse told us, that this Instrument would move
of itself, upon the motion of a Corresponding and
Sympathetick Instrument, which he had at home, and
that by means of this Instrument, my happy Master
had signified to him his approaching death; and that
after this signification, finding that this
Instrument remained without motion, he concluded my
Master was dead.

Last of all we came to the Books of Wisdom, which he
opened not;   near the said Books was placed a Box
of Gold, full of a most ponderous Powder of a deep
Scarlet colour, which Mr. Jesse smiling took and put
up.

Near to the Closet where we were, was another Closet
adjoyning, which we entered into, and there found
four large Chests full of small Ingots of most pure
Gold, out of which they gave me my Legacy of 6000
Golden Ducatoons in a double proportion. But Mr.
Jesse refused to take for himself any of the said
Gold; for he said, that those things which were
afore bequeathed to him, did fully content him, for
he was skill'd in my Masters Art, and therefore
ordered his part of the Gold to be bestowed upon
several poor Virgins, of kin to them, to make up
their Portions. I myself married one of these, and
had with her a good Portion out of the said Gold;
she embraced the Christian Religion, and is yet
alive.

Mr. Jesse packed up all his things, and carried them
home with him into Switzerland, though since that he
hath chose himself a quiet and well-tempered place
in the East-Indies, from whence he writ to me last

year, offering me to adopt my eldest Son, whom I
have accordingly sent to him.

During the time we were in the Closet, I saw strange
Miracles effected by the motions of the said
Instruments of Wisdom, which I neither can nor dare
set down in writing. Thus much, my intimate Friend,
I was willing you should know, more I cannot add.

Farewell.

FINIS.

A TREATISE OF MERCURY

And the PHILOSOPHERS STONE

By Sir GEORGE RIPLEY

I will, my dearest Son, instruct thee in this
Blessed Science, which was hid from the Wise of old,
to whom God was pleased to shew so much favour: Know
therefore, that our Matter is the chiefest of all
things in the Earth, and of the least estimation and
account, as will hereafter more plainly appear. For
if Water incorporate itself with Earth, the Water
will be lowest of all, and will (if it be not kept
down) with Fire, ascend higher; and thus it may be
seen, how Water will be the highest and lowest. Yet
true it is, that it is of least estimation, for in
our Earth and Water, and in that drossie Earth, you
may find some very pure and clear, which is our Seed
and fifth Essence, and then that foul and drossie
Earth is good for nothing else, and of no
estimation. But that Water, as I said, is the
chiefest, will appear many ways: Know, Son, that
without Water we cannot make Bread, nor anything
else, which God hath created in Nature; and hence
you may easily perceive, that Water is the first
Matter of all things which are born or generated in
the World: for certainly 'tis manifest unto thee,
that nothing grows or receiveth increase without the
four Elements; therefore whatsoever is Elementated
by the virtue of the four Elements, it must of
necessity be, that the original of all things that
are born or grow, should be of Water: Yet ought you
not to understand, this before spoken of Water, but
of that Water which is the Matter of all things, out
of which all Natural things are produced in their

kind. Know therefore, that first of all Air is engendred of Water; of Air, Fire; of Fire, Earth. Now will I more familiarly and friendly discourse with thee; I'le further manifest this Mystery unto thee by degrees, lest by too much hast it happen to us according to the Proverb, That he that makes too much hast, often-times comes home too late. Now therefore that I may satisfie thy desire, I will discourse of the first Matter, which Philosophers call, the fifth Essence, and many other Names they have for it, by which they may the more obscure it. In it for certain are four Elements, pure in their Exaltation: Know therefore, that if you would have the fifth Essence, Man, you must first have man, you must have nothing else of that Matter; and see that you observe this well. This I say that if you desire to have the Philosophers Stone, you must of necessity first have the fifth Essence of that same Stone, whether it be Mineral or Vegetative; Joyn therefore Species with Species, and Gems with Gems, and not the one without the other, nor anything contrary, which may be other than the Species or proper Gems; beware therefore of all that is not Essential: For of Bones, Stones cannot be made, neither do Cranes beget Geese; which if you will consider, you'll find the profit of it, by the help of Divine Grace by the assistance whereof let us farther proceed to speak of this blessed Water, which is called the Water of the Sun and Moon, hidden in the concavity of our Earth. Concerning which Earth know that all that is generated must of necessity have Male and Female from which action and passion arise, without which Generation never is. But you will certainly never receive profit from things differing in kinds. Notwithstanding, if you have this Water of the Sun and Moon, it will draw

other Bodies and Humours to its own kind, by the
help of the virtue and heat of the Sun and Moon, and
will make them perfect. As an Infant in the womb of
its Mother, decoction of temperate heat helping it,
turneth the Flowers into its nature and kind, that
is, into Flesh, Bloud, Bones, and Life, with the
other properties of a living Body, of which 'tis
needless to say any more. And hence you may
understand, that our Water changeth itself into a
perfect kind, with things of its own kind: For
first it will congeal itself into a substance like
Oyl; then it will change that Oyl, by the means of
temperate heat, into Gum; and lastly, by the help of
the perfect heat of the Sun, into a Stone. Now
therefore know, that out of one thing you have
three, that is, Oyl, Gum, and a Stone. Know also,
that when the Water is turned into Oyl, then you
have a perfect Spirit; when the Oyl is turned into
hard Gum, then you have a perfect Spirit and Soul;
and when the Spirit and Soul are turned into a
Stone, then you have a perfect Body, Soul and Spirit
together: which as it is called the Philosophers
Stone and Elixir, and a perfect Medicine of mans
Body; so also that which is leavened with its genus,
and the fifth Essence. Know, Son, that fifth
Essences are diverse, one whereof is to Humane
Bodies, another to Elixir, and to the imperfect
Bodies of Metals: For you must consider, that the
generation and growth of Metals, is not as the
growth of mans Body; for a genus agrees with its
genus, and a species with its species.   Moreover,
know that the first Matter of man, which begetteth
the Flesh, Bloud, Bones and Life, is a Spermatick
Humour, which causeth generation, through a vital
Spirit included therein: And when the Matter is
generated and congealed into a Body, extract thence

the fifth Essence of that Body, wherewith you may nourish the Body. Yet, Son, will I tell thee moreover, that Water, or Matter, or Seed whereof Man is begot, is not the augmenter of the Body. Know, Son, that if the Body be fed with its natural food, then its first Matter will be increased, and also the Body, (viz.) the first Matter in quality, and the Body in quantity; the first Matter is that which is called the fifth Essence. Yet know, Son, that the fifth Essence is one thing, and the Matter of Augmentation is another: and, as I said before, the increase of Metals, is not like the increase of mans Body. Although the fifth Essence, which causeth the augmentation of Metals, may be a fit Medicine for Humane Bodies; as also the fifth Essence, which causeth the augmentation of mans Body, may be a fit Medicine for the Bodies of Metals: and therefore, as before is said, the fifth Essence is one thing, and augmentation another. You see therefore for what reason our Water is called, the first Matter and Seed of Metals, viz. because of it all Metals are generated. Therefore you will have need of it in the beginning, middle, and end, for as much as it is the cause of all generation, because by its Congelation it is turned into all sorts of Metals, to wit, into the first Matter of the sorts. Thence it is called, the Seed of Metals, and the[51] Metallick Water of Life: because it affords Life and Bloud to sick and dead Metals, & joyneth in Matrimony the Red man with the White woman, that is, the Sun and the Moon. It is called also Virgins Milk; for as long as it is not joyned with the Sun and the Moon, nor with anything else, except only those which are of its

---

[51] So it is in the Latine, though perhaps the words should be Vita Metallica, that is, the Water of Metallick Life, that is, of the Life of Metals.

own kind, so long it may be called a Virgin. But
when it is joyned with a Male and Female, and
marrieth with them, then is it no longer a Virgin,
because it adhereth to them, and becomes one with
them to whom it is joyned, that is, with the Sun and
Moon, whom it joyns and is joyned with to
generation. But as long as it remains a Virgin, it
is called Virgins Milk, the Blessed Water, and the
Water of Life, and by many other Names.

And now, my Son, that I may say something of the
Philosophers Mercury, know that when thou hast put
thy Water of Life to the Red man, who is our
Magnesia, and to the White woman; whose name is
Albifica, and they shall all have been gathered
together into one, then you have the true
Philosophers Mercury. For after that in this manner
all is joyned with a Male and Female, then it is
called the Philosophers Mercury, the Philosophers
Water of Life, the Bloud of Man, his red Flesh, his
Body and Bones. Know therefore, that there are many
sorts of Milk, (Viz.) Virgins Milk, Womans Milk,
and also Mans Milk: For when first they are joyned
in one, and she is big, having conceived, then the
Infant must be nourished with Milk: But then you
may know that this Milk is not Virgins Milk, but
rather the man and the womans Milk, wherewith it is
always to be nourished, till it is grown to that
strength, that it may be brought up with stronger
and fuller food. That food which I mean is the
leavening of it, which gives it form, that it may
perform Virile work: For until the Infant, that is,
this our Stone, be formed and leavened with its
like, the Bloud of the green Dragon, and the red
Bloud of the red Dragon, whether it be the white
Stone or the red, it will never do a perfect work.

Know therefore, Son, that the first Water is that
Water Rebar, which God made of Nature, and it is the
cause of Generation, as I said before; but when
after the conjunction which ariseth from the
Marriage, it begets the Water of Life, and the
Philosophers Milk, with one of which, or both, you
must augment and feed your Stone perpetually.

Much more could I say to thee, Son, concerning this
first Matter, but let this suffice, that setting
aside impertinences of words, we may now, Divine
Grace favouring us, proceed to the practice itself
of the Philosophick Stone. See therefore, my Son,
that thou diligently puttest all these Matters
(which though they are three things, yet are they
but one only) in a Glass Vessel, and lettest them
quietly putrefie: then put an Alembick upon your
Vessel, and by distillation draw out all the Water,
which may be thence distilled. Try this first in
Maries Bath. Then place the Vessel in Ashes, and
make a gentle Fire for 12 hours: then take the
Matter out of the Vessel, grind it well by itself,
without the foresaid Water, then put it again into
the Vessel with Water, and stop the Vessel close.
Put it in the Bath for three days, and then distill
the Water as before in the Bath, and the Matter will
be more black than before. Do thus three times
over, and then grind it no more; but afterwards as
often as you distill it, so oft pour Water on the
top: but between each distillation give it so much
Fire for six hours or more till it become
indifferent dry; then pour Water on the top again,
and dissolve it again in the Bath under a blind
Alembick. Also in every distillation separate the
Flegm, by casting away six or seven drops of Water
in the beginning of each distillation. And

observing this order, cause it to drink its proper
Water, till it hath drank of it seven times its
weight which it had at the first. But then it will
be of a white colour, and so much the whiter, by how
much of the more of its own Water it hath drank.
This is White Elixir.

Moreover, this our Water is called Homogeneal, and
by many other names. Besides, know that this Water
and Matter generate as well the Red Stone, as the
White: Know also, when this first Matter is brought
to its compleat whiteness, then the end of one, is
the beginning of the other; that is, of the Red
Stone, which is our Red Magnesia, and Virgins Brass,
as we said at first: Son, see thou well understand
these words. Our Virgins Brass, is our Gold; Yet I
do not say, that all Brass is Gold: also our
Brass, is our live Brimstone; but all live
Brimstone, is not our live Brimstone: also
Quicksilver, is Mercury; but I do not say, that
common Quicksilver, is our Silver: as I said before,
that Water of Life which is our Seed and first
Matter, is our Mercury and our Spirit of Life, which
is extracted out of the bless Land of Aethiopia,
which is called Magnesia, and by many other names.
Besides, my Son, know that there is no perfect
generation, without corruption; for corruption
causeth cleanliness, and cleanliness corruption.
Consider therefore, Son, our dying poison, which
dyeth and is dyed perpetually; and this is our Body,
our Soul, and our Spirit, when they are joyned
together in one, and become one thing, which with
its parts ariseth also out of one thing, besides
which there is not any other, neither ever shall be.
Wherefore, my Son, great folly it is for anyone to
believe, that any other Medicine can be turned into

Gold or Silver; which Medicine will little profit
thee of itself, except it be mingled with a Body,
for then shall it perfect its work according to its
form to which it is born: For it is never born that
it of itself become a Body. Moreover, know that
there is as much difference between the first
Matter, which is called the Seed of Metals, and the
Medicine, as between the Medicine and Gold: For the
Seed will never be the Medicine without a Body,
neither will the Medicine ever be a Metal without a
Body. Much difference also there is between Elixir
and the Medicine, as between Masculine and Feminine
Seed, and also an Infant which is generated of those
in the Matrice. Now you may see, that the Seed is
one thing, and the Infant another; though they be
one and the same in kind, one thing, one operation,
the Vessel finally one, though it be called by
diverse names: For of a Man and Woman, is an Infant
born, when as yet the Man is one thing, and the
Woman another, though they be one and the same in
kind: which you ought to understand in our Stone.
But what I said before, that corruption is the cause
of generation, and of cleanliness, is true: For,
you must know, that everything in its first Matter
is corrupt and bitter, which corruption and
bitterness is called dying poison; which is the
cause of Life in all things, as will be sufficiently
manifest, if you with right reason do weigh the
Natures of things. Consider well, O Son, that when
Lucifer the Angel of Pride, first rebelled against
God, and prevaricated the Command of the most High,
be assured that this was made corrupt, bitter, and
harsh to him: No less was the fall and
prevarication of our first Parents Adam and Eve,
whom death and condemnation followed, made to them
corruption and bitterness, and likewise to us in

whom the same corruption is propagated. Many more like examples I could recite, if need were: But setting aside these, to come to what is proper to our discourse; consider well, that of all precious Fruits which grow out of the Earth, their first Matter is bitter and harsh, as still retaining some footstep of the former corruption and putrefaction; which bitterness, by the means of continual action of natural heat, is with great virtue turned into sweetness. Now therefore, Son, if thou wilt be ingenious, this little will suffice whereby to find out much more, and to perceive my meaning: Consider therefore well, Son, that according to the old Proverb.

He sweet deserves not, who no bitter tastes.

But now to speak something more of our Brass; know, that Brass signifies continuance, or continuing Water: But what is farther to be considered in the nature of the name of Brass, you may easily gather from its English Tetragrammate name, that is, its name consisting of four letters, to wit, B. R. A. S. First therefore, by B. is signified the Body of our Work, which is sweet and bitter, our Olive and our Brass continuing in its form: by R. is signified the Root of our Work, and the Spring of continuing Radical Humours, which is our Red Tincture, and Red Rose which purifieth all in its kind: A. signifies our Father Adam, who was the first man, out of whom was born the first woman Eve; whence you may understand, that therein is Male and Female. Know therefore, that our Brass is the beginning of our Work, our Gold and Olive, for it is the first Matter of Metals, as Man is the first of Man and Woman. S. signifies the Soul of our Life, and Spirit of Life, which God breathed into Adam, and all the Creatures; which Spirit is called the fifth Essence. Moreover, Son, by these four Letters, we may understand the four Elements, without which nothing is generated in Nature. They also signifie Sol and Lune, which are the causes of all Life, Generation, and augmentation of all things born in the World. In this name therefore of four Letters, consisteth our whole Work: For our Brass is Male and Female, of which ariseth he who is called begot. Therefore, Son, take good notice what is signified by our sweet Brass, what is called our Sandiver, or the Salt of our Nitre, or Nitre; what also by the Bloud of the Dragon, what Sol and Lune, our Mercury, and our Water of Life, and many other things, concerning which Philosophers have spoken darkly, and in Riddles. Know therefore, Son, that our first Matter is neither Gold, nor common Silver, nor is it of corrosives, or such outward things, which Denigrators groping in the dark now-a-days do use. Take heed therefore, Son, that by no means you admit anything contrary in kind; for be assured, that what a man shall have sowed, the same shall he reap. Moreover, know that when our Stone is compleated in its proper kind, then it will be a hard Stone, which will not easily be dissolved; yet if you add his Wife to him, he will be dissolved into Oyl, which is called the Philosophers Oyl, incombustible Oyl, and by many other names. Know therefore, Son, that there are divers leavenings, as well Corporal as Spiritual, (viz.) Corporal in quantity, and Spiritual in quality: Corporal leavening increaseth the weight and quantity of the Medicine, yet is not of so great power as the Medicine itself, as is Spiritual leavening; for it only encreaseth the Medicine in quantity, not in virtue: but Spiritual leavening increaseth it in both; and where the Corporal ruleth above an hundred, the Spiritual above a thousand. Moreover, as long as the Medicine is leavened by Spiritual qualities, so long it is called the Medicine; but when it is leavened with the Corporal substance, it is called Elixir. There is therefore a divers manner of leavening, and a difference between the Medicine, and the Elixir; for the Spiritual is one thing, the Corporal another. Know also, that as long as it is Spiritual leavening, it is liquid Oyl and Gum, which cannot conveniently be carried about from one place to another; but when it is Corporal, then it will be a Stone which you may carry about in your pocket. Now therefore you see what is the difference between the Medicine, and the Elixir; nor is the difference less between Elixir, and Gold and Silver, for Gold and Silver are difficult to melt, but Elixir not so, for it easily dissolves at the flame of a Candle: thence you may easily perceive, how various the differences of our composition and temperament are. Lastly, that we may say something concerning their food and drink, know that their food is of airy Stones, and their drink is drawn out of two perfect Bodies, namely, out of the Sun and Moon; the drink that is drawn out of the Sun is called liquid Gold, (or Potable, that is, that may be drank;) but that out of the Moon, is called Virgins Milk, Now, Son, we have discoursed plainly enough with thee, if Divine Grace be not wanting to thee; for that drink that is drawn out of the Sun, is red, but that out of the Moon, is white; and therefore one is called liquid Gold, but the other Virgins Milk; one is Masculine, the other Feminine, though both arise out of one Image, and one kind. Son, ponder my words, otherwise if thou wanderest in the dark, that evil

befalls thee from defect of light: See therefore
that thou beest diligent in turning the Philosophick
Wheel, that thou mayst make Water out of Earth, Air
out of Water, Fire out of Air, and Earth out of
Fire, and all this out of one Image and Root, that
is out of its own proper kind, and natural food
wherewith its Life may be cherished without end. He
who hath understanding, let him understand.
Glory to God Omnipotent.

FINIS.

Colours to be Observed in the Operation of the Great Work

You must expect to have it exceeding Black, within 40 days after you have put your Composition into the Glass over the Fire; if it be not black, proceed no further, for it is unrecoverable: it must be as black as the Ravens Head, and must continue a long time, and not utterly to lose it during five months.

If it be Orange colour, or half Red, within some small time after you have begun your Work, without doubt your Fire is too hot; for these are tokens that you have burnt the Radical humour and vivacity of the Stone.

Know ye not, that you may have Black of anything mixed or compounded together with moisture: But you must have Black which must come and proceed of perfect Metaline Bodies, by a real Putrefaction, and to continue a long time.

As for the colours of Blew and Yellow, they signifie that the Solution and Putrefaction is not yet perfectly finished, and that the colours of our Mercury are not yet well mingled with the rest. The Black aforesaid is an evident sign, that in the beginning the Matter and Composition doth begin to purge itself, and to dissolve into small Powder, less than the Motes in the Sun; or a glutinous Water, which feeling the heat, will ascend and descend in the Glass: at length it will thicken and congeal, and become like Pitch, exceeding Black; in the end it will become a Body, and Earth which some call Terra Saetida; for then by reason of the perfect Putrefaction, it will have a scent or stink like unto Graves new opened, wherein the Bodies are not thorowly consumed. Hermes doth call it Terra Soliis, but the proper name is Leton, which must be blanched and made white.

This blackness doth manifest a Conjunction of the Male and Female, or rather of the four Elements. Orange colour then doth shew that the Body hath not yet had sufficient digestion, and that the humidity (whereof the colours of Black, Blew, and Azure do come) is but half overcome by the dryness.

When dryness doth predominate, then all will be white Powder: It first beginneth to whiten round about the outward sides of the Glass; the Ludus Philosophorum doth say, that the first sign of perfect whiteness, is the appearing of a little hoary circle passing upon the Head shewing itself round about the Matter on the outward sides of the Glass, in a kind of Citrine colour.

THESAURUS,

Sive Medicina Aurea

A plain and true
DESCRIPTION
OF THE
Treasure of Treasures, OR THE
Golden Medicine

Many and great are the Secrets of Nature, and concerning them and the way to attain them, the wise Philosophers have writ much, but in a very dark and Aenigmatical stile, so that very few are those that attain to anything of their desires by them; but on the contrary, after much time, labour, and cost in vain expended in the search of them, are forced to give over at last, and surcease their further inquiry, and instead of the desired satisfaction, conclude from their lost labours, that the Books of the Philosophers are only fabulous, and writ to deceive the unwary, and those that thirst after so great a Treasure. But I vow unto thee by Almighty God, that what they have wrote is a real Truth, though delivered in so dark and dubious a way, that few are able to understand and receive benefit from them. I do therefore attest the truth of their Medicines, as well for the transmutation of the baser and imperfect Metals into Gold and Silver, as for the benefit of Humane Bodies, and healing all Bodily Diseases, till God calls the Soul; and this above all the Medicines of Galen and Hippocrates. But because many great lovers of Knowledge, and this Science, have so often failed of obtaining the end of their desires in these Mysteries of Nature, and

not only failed as to the Accomplishment of their
desires in full, but also come short of knowing the
principal subject, and ground of Philosophical
Secrets: I have therefore thought fit to help them
by this small Script, as much as I may, and save
them the troubles of that search; and by telling
them in plain terms and words the true Matter,
enlighten and encourage their dubious minds to the
farther search after what they desire: for let them
assure themselves, it is no small advantage to be
assured of the true Matter and groundwork, or Basis
of so great Arcana's and hereon great and
innumerable blessings do depend. I do therefore
most faithfully assure thee, that the true Subject
of this Art is Quick-silver, in a double manner,
viz. either Quick-silver Natural, or Quick-silver of
Bodies, viz. the Bodies, of Sol or Luna, reduced to
Mercury vive; for many and strange things may be
performed by either, singly of themselves, or else
conjoyned. The conjunction of the Mercury of Gold
or Silver, with the common Mercury; or the Bodies,
or the Oyl of Gold and Silver, dissolved in the Aqua
Mercurii, doth much hasten the operation of
Medicines for Metals: But we need not (as
absolutely necessary) any more than the common
Mercury or Quick-silver, dissolved lightly, either
for Elixirs or precious Stones; only small Natural
Stones must be dissolved in the Mercurial Water, so
shall you have such Stones again as you dissolve,
and those of what bigness you please, far exceeding
Natural ones.

The next great Secret of Philosophers, is the
preparation of common Quick-silver or Mercury; for
common Mercury, as Nature produceth it, is not fit
for such operations, nor can they any way be

performed by it: for our Mercury is not the common
Mercury or Quick-silver, but is made of it, by a
true Philosophical skill; it is not the white
Mercury or Quick-silver, but its subtle, spiritual,
airie and fiery parts, the earthy and watery being
prudently separated. For the manifestation of our
Mercury, the true Mercury of Philosophers, first
prepare the common Mercury by a due Philosophical
preparation, until thou hast separated and purged
him from his two extremes or excrements, Earth and
Water: dissolve it then, after its purification,
into a Milkie, Crystalline, and Silver Liquor or
Water, which in three or four months is to be done:
being once dissolved, thou mayst ever after dissolve
more and more Mercury in fully fourty days, for
Mercury once dissolved, dissolveth it self ever
after infinitely. And having dissolved it, distill
it perfectly, until it have no Faeces in the
Cornuae; after Distillation, bring it again to
Putrefaction, and when it is blackish, distill it
again: so shalt thou have two Oyls, a white Silver
Oyl, and thickish, and at last a very red or Bloud-
like Oyl, which is the Element of Fire. The white
Oyl serveth for Multiplication, or multiplying the
white Elixir, and for the making of all precious
Stones, by dissolving of small precious Stones in
it, for it will presently dissolve them: then in a
gentle heat of Ashes congeal them again, and they
far exceed any Natural ones, both in lustre, and
virtue, and hardness. The red Oyl is for the
multiplying the red Elixir, even to an infinite
height in projection; which when it is by often
multiplying or multiplication, brought to a fixed
Oyl, then thou mayst do several Magical, yet Natural
and strange Operations by it.

To make the Elixirs thou must proceed thus: When
thou hast dissolved rightly the common Mercury,
which cannot be done before it be duly prepared for
such a Philosophical dissolution, when it is
dissolved into a Milkie, Silver, Crystalline Liquor,
it will in the distillation leave some Faeces, in
which remains its more fixt part or Salt, which thou
must warily and wisely, after a gentle and
Philosophical Calcination, extract and purifie to
the highest Purification, by which means it will be
very white and clean: then take seven ounces of the
white Mercurial Oyl, and dissolve in it as much of
this Salt as it will dissolve, until it will
dissolve no more: having so done, put thy Liquor
into a Philosophical Egg, sealing it Hermetically,
and by due degrees of Fire congeal and fix it: being
fixed, it is the white Medicine, which fermented
with Silver, may be cast upon purged Venus, which it
will transmute into most fine Silver: multiply it
with the white Oyl, &c. If thou wouldst have it be
a red Elixir, put to it some of the red Oyl, and by
requisite degrees of heat congeal and fix it as
before: ferment it with Gold; multiply it by the
red Oyl, and the aforesaid white Salt dissolve in
it: dissolve it, congeal and fix it often, until
it will congeal no more, so will it remain an oyl,
which in its projection is almost infinite,
Endeavour not to multiply it any farther, for fear
thou losest it; it is then so fiery, that it will
vanish out through the Glass reddish, or Rubie-like.
Make projection with it on what Metal thou wilt and
thou shalt have most fine Gold, better than the
Natural Gold. Laus Deo, &c.

Mercurius albus & rubeus ex Mercurio vulgi (per Se)
Soluto fieri potest tanquam ex Mercurio Universali:
Age Deo & mihi gratias. E.B. &c.

TRACTATUS DELAPIDE,

Manna Benedicto, &c.

In this Book thou hast a most faithful and plain Manuduction to the greatest and most noble Secret of Nature: Enjoy them in silence; bless God, and do good unto thy Neighbour and Successor, as I do hereby to thee, thou finder of this Book.

I have resolved with myself to write this short Treatise, having been not only an eye-witness, but also an actor of such high Mysteries of Nature, as the World is not worthy of, and the Wise of the World do scarce believe. Which discourse may be of singular use to such as God shall please (out of his infinite mercy) to bestow the knowledge of this Stone upon, to make the Stone of the Wise men so called, or the Philosophers Stone; which shall be of much use and benefit to those who are not yet capable of making the Stone itself, for it shall illuminate the understanding of all that read it, more than all the Books they shall read: For it shall set down the Basis and Foundation wherein the wisdom of all the Philosophers doth lye, (I except none;) yet not so as to name that (which no man durst) in so plain words, that every fool or lewd fellow may understand it, as he may his A.B.C. when he reads it, for that were to make my self accurs'd. Whosoever thou be that readest this, let me advise thee rather to fix thy mind and Soul on God, in keeping his Commandments, than upon the love of this Art; which although it be the only, nay all the wisdom of the World, yet doth it come short of the Divine Wisdom of the Soul, which is the love of God

in keeping his Commandments. Yet let me tell thee,
he that shall have the blessing to make the Stone,
and find this writing, he shall see such Mysteries
in Nature, as shall make him of a wicked, a good
man, or else a very Devil incarnate. But I am
perswaded it shall never be permitted to come to the
hands of any but whom God knoweth fit for it, and
such as shall never abuse it. Hast thou been
covetous, prophane? Be meek and holy, and serve in
all humility thy most glorious Creator; if thou
resolve not to do this, thou dost but wash an
Aethiopian white, and shalt waste an Earthly Estate,
hoping to attain this Science. There is no Human
Art or Wit can snatch it from the Almighty's hand;
nor was it ever, nor I am perswaded ever shall be,
given but to such as shall be of upright hearts.
Remember what King David saith, The Fear of the
Lord, is the beginning of wisdom, a good
understanding have they that do thereafter: and so
if thou think to attain this wisdom, which is the
top of all wisdom, and indeed angelical wisdom and
yet dost not fear the Lord; thou dost give King
David, and in him the Holy Spirit the Lye, which be
far from every Christian heart. But let me conclude
my preface with this: If God bless thee with the
Stone, and thou have the enjoyment of this little
Script, and dost make that use of it that here is
set down, thou shalt see that which is not fit to be
written, yet I have set down in part what thou shalt
see hereafter: as thou shalt read, pray and study;
pray with a faithful and earnest heart, study with
an honest heart, and leave the issue to God, to whom
be Glory.   *Amen.*

The folly of the Students in this Noble Science and
Art, is this; they set their minds and intentions on

nothing but making of Gold and Silver, and so they fall into this errour, that Gold and Silver must be the ground-work of this goodly piece; but that is false: yet will I not now stand to disprove it, for that were tedious; it is sufficient that I vow upon my Soul, it is not so, nor any such matter: yet it is true, that it hath a true Golden and Metalline Nature. But to proceed, briefly know, that the changing of imperfect Metals into Gold and Silver, as it is the chief intent of the Alchymists, so it was scarce any intent at all of the Ancient Philosophers; and although it be to be done by this Art, yet it is but a part, and indeed the least part of the benefit that cometh by the Art: yet I deny not but the possession of Gold and Silver is a great blessing, especially got in this way, because it freeth a man from want, and being beholden to others; as also that a man may do good to others, to the poor and oppressed; nay it is a happiness in this World to possess much, but yet I affirm it the least happiness that cometh by the Philosophers Stone, if the full use thereof be known. Gold and Silver are goodly things, and the enjoyment of them very delightful to covetous and wicked-minded men, who do not trust in God, and know him as they ought; but a true searcher of this Wisdom, is content, as the Apostle saith, with meat, drink and cloaths, viz. a competency. I have a little exceeded in my exclamation against Riches, because I know it befits not a wise man to love them: when thou hast read all that I have set down, thou wilt not value Wealth, as thou wilt other Knowledge herein set down and contained; for by the full knowledge of it, the whole wisdom of Nature is to be grasped and embraced; yea not only infinite Wealth, and perfect Health, (a far greater blessing than Wealth) but also the knowledge of all Animals, Vegetables, Minerals, the Radix and Root of all which, is the true Root of all Philosophy; nay more, of all the seven Liberal Sciences, which in their full perfection are to be known by the knowledge of this Art, and without it not one can be perfected; nay more, the artificial making of all precious Stones, better than the Natural, and of what greatness you please, as Rubies, Carbuncles, Diamonds, Jacinths, Pearls, Topases, Saphirs, Emeralds, &c. But this is not all, for by the perfection of this Art, which very few have attained unto, all Natural Magick may be known, all that Spirits can do (except velocity) may be performed by a true Philosopher, though to ignorant men it seemeth supernatural; all that is natural may be done by this Art, wicked Spirits may be commanded and driven away; in a word, whatever is sublunary may be done by it. All these things were known to Adam in his Innocency, who had this Art in the highest perfection. This man, our first Father, was in his inward parts, or internal man, made according to the similitude of God; and to tell the plain truth (which when thou hast tryed what is here set down, thou shalt know it to be so) was of the same Substance and Matter, that the Angels were made of, I mean the blessed Angels. The Soul of man is an Angel, and so was called the Son of God; but for his Body and Spirit, whence that came, and what they are, I will set down by and by that which concerneth that. Man was the Son of the great World, or Macrocosm, and participateth of all the influences and virtues of the superiour and inferiour Worlds, yea of all Creatures good and bad, and that for this cause, because he was made of that very Matter and Chaos whereof all the World was made, and all the Creatures in it: which is a most high Mystery to

understand, and must, nay is altogether necessary to
be known of him that expecteth good from this Art,
being the ground of the wisdom thereof. Foolish
men, nay they that the World holds for great
Doctors, say and tell it for truth, that God made
Man of a piece of Mud, or clay, or Dust of the
Earth; which is false, it was no such Matter, but a
quintessential Matter which is called Earth, but is
no Earth. The Fall of Man deprived all things, yea
all the Creatures suffer'd in it, and himself most
of all; for as soon as he had sinned himself, and
his Wife, his Wife first, both of them turned into
Monsters in respect of what they were in their
Innocency. Adam had another Body before his Fall,
than what he had after; and so far different, that
if we should behold Adam as he was in his Innocency,
we should admire the glory of him, and tremble at
the sight of him, as at the sight of an Angel. I
say nothing of the Body of our blessed Saviour, save
only this; such a Body as he brought from Heaven
with him, such a Body shall we arise with, and with
such Bodies shall our Souls be endowed with flesh
and bloud; otherwise men should not differ from
Angels, for this flesh and bloud is put upon us by
the Holy Ghost, that is, by Regeneration: nor doth
this cross Sacred Writ, if it were rightly
understood. I speak nothing likewise of our blessed
Lady, what Body she had; but when thou shalt have
examined what I shall set down, then wilt thou find
what I say is true, and understand them in a plain
manner: but I forbear to speak of those Mysteries,
known to so few; he that liveth according to the
written Word of God, shall be saved; but he that
liveth to be blessed with this Art, shall glorifie
his Creator, and know him more than any man can do;
but before the end of the World, all will be known.

But to my former purpose: Man, the Microcosm, or little World, from the Astres or Stars received Spirit, from the great World his Body, and from God immediately his Soul, so here is an illucidation of the blessed Trinity; of these three Man consisteth, of these he is compounded, thus he had his production from the World. For what concerns his Body, or Humane part, let us now (as fully as we may) say something of the production of the great World out of nothing: when there was neither time nor place, did God create a certain Chaos, invisible, intangible, which the Philosophers called Hyle, or the most remote Matter; out of this he made an Extract, or second Matter or Chaos, which the Philosophers know, not by speculation, but by sense: that Matter was and is visible and tangible, in which were and are all the Seeds and Forms of all the Creatures, superiour and inferiour, that ever were made: from this God divided the four Elements; in a word, did make all things celestial and terrestrial, the Angels, Sun, Moon, and Stars. The knowledge and practice of the Philosophers upon this Chaos, brought them to the knowledge of all wisdom, and from hence (next God) seek thou and find all wisdom.   This is not a fancy or conceit that I tell thee, but what I know and have proved; it is such a thing and substance, as with the bare knowledge of it, makes thee know the generation and preservation of all things, and yet this Chaos is since the Fall likewise corrupted. Thus briefly have I discoursed of such things as perhapse thou dost not believe, or never heardst of before; but if thou be'st ordained to know this Science, I have trod the path for thee, but I fear thou understandest me not: yet have I said more than wisdom would I should, but I know it shall be to his

good only, to whom God appoints it. My intent is,
for certain reason that I have, not to prate too
much of the Matter, which yet is but only one thing,
already too plainly described; nor of the
Preparation, by which means it is to be done, which
is the second and greatest Secret: But I have
constituted these lines for the good of him that
shall make the Stone, if it fall into the hands of
such a one; for to him it shall shew and set down in
plain terms, as plain as possibly my Pen can write
to the very letter, such Magical and Natural uses of
it, as many that have had it never knew nor heard
of; and such as when I beheld them, made my knees to
tremble, and my heart to shake and I to stand amazed
at the sight of them. I do therefore charge thee,
whosoever thou be that shalt be blessed with the
enjoyment of this Treatise, that as thou wilt answer
the contrary at the great day, thou let no man see
it, but him that hath the Stone perfect; for if thou
shalt meet with such a one, (which is hard to do)
and that he hath brought it to the full perfection,
thou by imparting such Magical and Physical things,
and other rare Secrets which are here set down, and
by the Stone to be done, he shall not only give thee
Gold sufficient, but also shall shew thee the true
and right way, and the Matter with all things
belonging to it, to make it full and perfect: for
let me assure thee, I have known many that have had
it, that never knew more than the bare transmutation
of Metals; and by the Books of the Philosophers it
appears, that some of them, (nay more than that)
many of them have kill'd themselves by taking it for
want of the knowledge of the use of it. Never doubt
therefore, but thou shalt obtain what thou wilt of
him that hath it, by demonstrating the truth of what
I here write; therefore again and again I charge

thee not to part with it, nor to tell any man of it, although none can make use of it, but he that hath the Stone in the highest degree of perfection. And I will now shew thee the several uses of it: The first, for Health, and the manner of how to use it; the second is for multiplication, which cannot be done without a Master; Thirdly, the making of all manner of precious Stones artifically, better than the Natural; Fourthly to turn all Metals into running Quick-silver; Fifthly, several Magical Operations of several kinds, which are past belief, till thou seest them, and which indeed are above all the rest. And here I promise, that I will in such plain words set down what I have intended, that thou canst not in doing err, or do amiss, provided thou have the Stone both red and white, although there be more works out of it than I dare set down; and Indeed Angelical wisdom is attained by it. But I proceed.

For Health, the use of it thus.

In the use of this Medicine, many great Philosophers themselves, after they obtained this wonderful blessing, desiring to have perfect Health, have been so bold as to take a certain quantity of it, some no more than a quarter of a grain, some less, some more, but all that did so with it, instead of Health, took Death itself; for there is no small skill to it for Medicine, though every fool think if he had it, he could cure all diseases, and himself too, and set the Elements at unity, which few men have known, neither is there but one way to it with safety; if this be not known, more hurt than good may be received by it. For the method of Health, it is thus: Take the quantity of four grains, I do not mean the grains of Wheat, or Barley grains or corns, but four grains of Gold weight, and dissolve them in a pint of White or Rhenish Wine, but in no hot Wine, as Sack, &c. put it into a great clean Glass, and instantly it will colour all the Wine almost as red as itself was, which is the highest red in the World: let it stand so, close covered from dust, four days, for in respect it is an Oylie substance, it will not presently dissolve in Wine; then add to this a pint more by degrees, until it be not so red, stirring it with a clean stick of Wood, not of Metal, nor Glass, and so continue the pouring on of fresh Wine, until it be just of the colour of Gold, which is a shining yellow. Beware there be no redness in it; for so long as there is any redness in it, it is not sufficiently dilated, but will fire the Body, and exhaust the Spirits: neither is it sufficiently brought to yellow, until the Wine have round about the sides a ring like Hair, of a whitish film, which will shew itself plain when well dissolved, if it stand but four hours quiet. As soon as you see that whitish film, then let it run through a clean linen Cloth, or Paper, so the white film will stay behind and look like a pearl on the paper: and all the rest will be yellow like Gold. This is the token of truth, that you cannot wrong yourself by this Liquor; and without this token, it will be either too weak, or so strong that it will fire the Body. Know this to be a rare Secret. Of this Golden Water, let the party (of what disease soever he be sick of) take each morning a good large spoonful, and it shall expel the disease whatsoever it be, by a gentle sweat; for it purgeth not, nor vomiteth, nor sweateth so much as to make faint, but to corroborate: I say, it strengthens the party; and if the disease be of many years continuance, or a Chronical disease, it will then be perhaps twelve days, otherwise but twenty four hours, or two or three days at most. Thus it must be used for all diseases internal: But for all external diseases, as Ulcers, Scabs, Botches, Scores, Fistual's, Noli me tangere's &c. the place must be anointed with the Oyl of the Stone itself, not dilated in Wine; and after this manner it must be done nine or ten days, and be it whatsoever it will, it will cure all outward and inward diseases. And more than this, whosoever carries this Stone about him, no evil Spirit can or will stay in the place; nay bringing or giving it to a party possessed, it drives away and expels the evil Spirits: for it is a Quintessence, and there is no corruptible thing in it; and where the Elements are not corrupt, no Devil can stay or abide, for he is the corruption of the Elements. This Medicine taken nine days as aforesaid, and the Temples of the Head anointed with the Oyl of the Stone each day in the morning, it will make a man as light as if he could flie, and his Body so aireal it is not to be credited, but by him that hath experienced it. These most admirable qualities it hath, perfect health it giveth, till God calls for the Soul; and perfect knowledge it giveth, (if the use be known:) but even this part hath been known but to a few that have made it, for it is a Divine, and as it were an Angelical Medicine. The white is not to be used for any disease but Madness, in the same proportion, and way or preparation that the red Stone was: And so I proceed to the second which is Multiplication.

The way to Multiply

Many have made the Stone both white and red, that
never knew how to multiply it, for the white Stone
will be red, by continuing it in the external
natural Fire; but never make projection higher than
one upon ten, neither white, nor the red: few have
known this, for if they be not armed rightly, it
will kill them; but do thus, and thou shalt multiply
it infinitely, that it shall not congeal to Powder
any more. When thou hast made the Mercury of the
Philosophers, (which in 40 days is to be done) a
Water it is, and no Water, clear as the Heavens,
then as thou didst make it, reduce it back again
into Putrefaction, E.F. which it will quickly do in
a Body with a blind head, and never put into it
above twelve ounces, and lute it with such lute as I
will direct here-under, for in a Glass nipt up it
will not work. When it is like Pitch, take out thy
Glass, and remove it to a common fire of Ashes in a
Furnace, and when thy Glass hath stood cold 24
hours, arm thy self thus: Make thee a Case for thy
head and face with Hog-skin, lined with Cotton, and
before thy face have Spectacles of Glass, and from
thy mouth let go a large Tunnel of Glass, covered
with Leather, and let it be tyed under thy Girdle
and touch thy Ancles; let the bore of the Glass be
as big as a Walnut, and tye the Hyde of Hog-skin
fast about thy Neck under thy Chin, but so as thou
be sure no Air come in there, to which purpose lap
it over with more Hog-skin, basted with Laten: and
thus art thou well armed, for otherwise it would
kill thee. Thus armed, take off thy Blind Head,
and put on a distilling Head, and a Receiver long
and large: lute the Receiver and joynts of the Head
with this lute, (viz) to one ounce of Powder of

Egg-shells, calcined 24 hours, and ground like Meal,
take two ounces of Enamel, such as the Goldsmiths
use; grind that with the Eggshells, and add the
white of an Egg to make it into paste, but the white
must be well beaten first: then smear this upon
Bladders made supple, and herewith anoint the joynts
of the Receiver three times double; let it dry 24
hours. Put thy Glass in Ashes but six fingers above
the Matter thus putrified, and let the head of the
Glass be very cold, and with a gentle heat you shall
see a white fume arise, and make all the head of the
Receiver like Milk; increase easily that Fire, till
no more will come, then let all cool, and these
white fumes settle to a white Water thickish; this
is that white Mercury to multiply the white Stone:
then put a new Receiver, luted as before; put in as
many Coals as the Furnace will hold or bear, till
the Pot be red hot, then shall you see the yellowish
fume arise, and instantly will it increase redder
and redder. Continue the Fire until an Oyl come
redder than Bloud into the Receiver, and it will be
also thickish; this is the red Mercury, wherewith
the red Stone must be multiplied: each of these
must be new rectified, in a new Body and Head, till
they let no Faeces, which will be in seven times,
and then stop them close with the same lute till you
use them; and when they are cold, they are white and
red Oyl flowing in the bottom, which will melt with
an easie Fire, and being cold, be as a Salt: these
are the three principles of Salt, Sulphur, and
Mercury, a plain Elucidation of the blessed Trinity.
Now when the white Stone is made, it will not melt,
but is like white Sand, but impalpable, and will
tinge no Body but Venus into Luna. To three parts
of the white Stone, take one part of white Mercury
rectified, but first dissolve in that white Mercury

one third part of white Salt; then imbibe the white
Stone, which will presently take it, and be like
Pap; then close your Egg (for so is your Glass
Multiplier) with the aforesaid lute, and set it in
your first Fire, H. E. I. E. F. and it will in 40
days putrifie, and pass all the colours, and be
white fixed, and project one part upon an hundred:
repeat that with more white Mercury, as before,
keeping the same proportion and the same Fire, and
it will multiply each time ten, at the third time it
will be a thousand, then ten thousand, then a
hundred thousand, so you may bring it to a white
Oyl, like the Moon pale in the dark; then it will
multiply no more, neither will any Glass hold it.
If you make projection with the white Stone, then
melt fine Silver a tenth part, then cast in the
Stone; keep it 24 hours melting, and this is
Fermentation. The first time the white goeth only
upon Venus, the second time upon all Bodies, the
third time upon common Mercury, and then it is
Elixir of Spirits. As you did with the white, so do
with the red exactly; but take the red Mercury, and
white Salt, and so that goeth one upon ten on Luna
the first time, the second upon an hundred, and so
to an infinity; and so it will be red Oyl like a
Carbuncle, and will shine in the darkest night with
admirable splendor, and from it will flie all evil
Spirits. And this they must have, before they cure
all diseases, and give that exaltation to man, to
make such Magical words as I shall set down.

To Make Stones

Having made Mercury of the Philosophers, and out of
it the two Mercuries white and red, if thou wilt of
small Pearls make great and Oriental ones, do thus:

Take white Seed Pearls, and dissolve them in the white Water, which will instantly of itself dissolve them: when it is like Pap, that thou mayst work them with thy hands, make it into pearls; and have a round mold of pure Silver, put thy Pap into this mould, but first anoint thy mould with the white Stone, which is an Oyl: when they have layn three or four days, open it, and lay the Pearls in the Sun, but not too hot, and they will grow hard, and more orient than any Natural Ones.

To make Diamonds.

Take the whitest Flint Stone you can get, beat off the outside, and dissolve the rest, as much as thou wilt, in the white Water: when it is dissolved to clear Water, not to Pap, put it into a little Vial, stop it close, and set it in warm Ashes, and in twelve days it will congeal to a hard gray Stone: then increase the Fire, that the Glass may be red hot, then let it cool; take it out, and it will be like a flint; but polish it, and thou never sawst such a sparkling Diamond, nor so hard: but it will be better if thou dissolve little Diamonds. All Stones that you dissolve in the white Water, the same colour they were of, the same will they be of; but for Rubies and Carbuncles, and all red Stones, they are made of the red Mercury, and of Crystal; and for a Carbuncle, you must add to ten parts of Crystal, dissolved in the white Mercury, one part of the red Stone brought to the highest, and so as before congeal it with Fire, and being polished it shineth in the dark beyond all whatever.

To turn Metals into Quick-Silver.

Do thus in the operation of the Stone white and red: when the white Stone first is made, never after thou shalt perceive lye under the glistering Powder, but thou canst not perceive it, till thou tak'st out the Glass; a grayish light subtle Powder, and the proportion is about 1/10 of the Matter put in. Put any Metal what thou wilt into a Silver Bason, (except Gold or Silver) and make a Plate as thick as you will, and in the middle a hole like a Barley corn, and in that hole put the Powder; to each pound of the Metal, six grains of the Powder, and no more; and as soon as it is hot, the Powder will eat into the Metal, and turn it all into Quick-silver: then pour it into Water, and the scruff will remain behind. For Gold and Silver, hold them so used over the Fire, till they turn to Quick-silver, then hold them over a wooden dish; this Powder is the Terra damnata of the Stone. Now I will shew thee that which is above all, certain Magical operations with the Stone, such as thou wilt wonder at, and bless thy Creator, when thou shalt see them: Wonders above wonders, nor wilt thou believe till thou hast done it.

The Creation.

Take Ordinary Rain-water a good quantity, ten gallons at the least, stop it up close in Glasses fourty days at least, and it will stink, and set a Faeces[52] at the bottom; pour off the clear, and set it in a Vessel of Wood, made round like a Ball, cut off in the midst, and fill the Vessel one third part full of it, and set it in the Sun at Noon-day, in a

---

[52] Gur.

private place: that done, take one drop of the red blessed Stone,[53] and let it fall into the midst of the Water, and presently thou shalt see a mist and thick darkness upon the face of the Water, as it was in the first Creation: then put into it two drops more, and thou shalt See the second light come out of the first darkness, or rather light come out [of] darkness; and then by degrees each half quarter of an hour put in three, four, five, six drops and then no more, and thou shalt see appear before thy face on the surface of the Water, by degrees one thing after another, all things that God did create in six days, and the manner of it, and Secrets not to be spoken of or revealed; which to reveal I have no power, nor strength, nor dare set down. Be on thy knees from the beginning of this operation, let thine eyes be judge, for thus was the World created: You cannot but tremble when you shall see it: let all alone, it will vanish away in half an hour after it begins. By this you shall know and see plainly those Mysteries of Divinity, which now you are ignorant of as a Child, although you thought yourself a wise man, and that you did understand Moses his Writings of the Creation; but I say no more. You will now see what Body Adam and Eve had before their Fall, and what after their Fall; what the Serpent was, what the Tree was, and what Fruit they did eat; where and what Paradice is, and what it was, you will know: What Bodies the Just shall rise in, not these we received from Adam, but that flesh and bloud which is born and begotten in us by the Holy Ghost and Water, such as our blessed Lord brought from Heaven. But I have done.

---

[53] In the translation by Jung, "red blessed stone" is rendered "consecrated wine".

The Heavens.

You shall take seven pieces of Metals, of each of
the Metals named after the Planets, and on every one
of them you shall stamp the sign or character of the
Planets; in the House of the Planet, and let each
piece be as big as a Rosenoble, only let Mercury be
of a quarter of an ounce, and no impression on it:
Then put them (as they stand in order in the
Firmament) into a Crucible, and close all the
windows in the Chamber, and let it be dark, and in
the midst of the Chamber; then melt them all
together, and drop in seven drops of the blessed
Stone, and presently (out of the Crucible will come
a fiery flame, and spread itself round about the
whole Chamber; fear it not, it will not hurt you)
the whole Chamber will shine brighter than the Sun
and Moon, and you shall see over your head the whole
Firmament, as it is above the Starrie Skie; and the
Sun, Moon and Planets will go all round in their
course, just as it is in the Heavens. Let it cease
of itself; in a quarter of an hour it is gone to its
proper place.

Fellowship.

More than this, if thou take the Stone each Full
Moon, when it is over the Horizon where thou art,
and go apart in a Garden, and take some of the clear
Rain-water, as thou didst in the first operation,
and drop of thy white Stone as thou didst of thy
red, and there shall presently even to the Orb of
the Moon ascend Exhalations in a strange manner; and
if thou observe this every month at the due time,
there is no Philosopher in the Horizon where thou
livest, that hath the knowledge of the Stone, with

the use of it, but at the same time goeth out and looketh East and West, North and South, and finding such an Apparition, (as he soon seeth it) he knoweth it is done by some Artist or other, that desireth acquaintance with those that have the same Art, and will presently in the same manner answer thee, when thine is done: thus shalt thou know all that have the use of the Stone. To meet with thy fellow Philosophers, do thus anoint thy temples with the white Stone that night, and earnestly pray to know what that party is: lay under thy head three Bay-leaves newly gathered, and fix thy Imagination upon thy desire to know him, so repose thy self to sleep; and when thou dost awake, thou wilt presently remember thy Vision, as the Person, his Name, and the place of his abode: if thou go not to him, he will come to thee, for perhaps he thinks thou dost not know this Secret. The reason why this should be thus, is this; the universal Spirit of the Air, which is inclosed in the Stone, causeth it. Thus mayst thou accompany thy self with all the wise men in the World, who shall appear unto thee rather Beggars, than Rich men, and perhaps can teach thee more than I can, or have done by this; for indeed all things that are Natural are done by it, a Volume would hardly contain them: As to command and converse with Spirits, which I forbear to set down, I mean good Spirits, is not this Angelical wisdom to know these things? Astronomy, Astrology, and all the Arts of the Mathematicks, are easily known in their perfection, this being done that I have told thee; nor is Scholarship required, it is the gift of God. You must know, before you do these things, you must take the Stone nine days, as I prescribed first, and it will make thee have an Angelical understanding; thou wilt despise the World, and all in it; then thou wilt know how to serve God, and understand the Scriptures.

I have written that which was never writ before; think whether they be not Secrets and Arcana's and whether thou ought'st to shew this or not to any man, but to him that hath the Stone. I have now done, charging thee to have a care of this Writing, commanding thee to serve God; for without thou do that, thou wilt never have good of this Art: Serve him in Spirit and Truth, and so to God I leave thee, to direct thee in his ways.

Glory to God in the Highest.
Amen.

Nicolas Flammell's Summary of Philosophy.

He that desireth to know how Metals are transmuted, he must know from what Matter they are, and how they be formed in their Minerals; and lest herein we err, we must see and observe the transmutations as in the Veins of the Earth. Minerals out of the Earth may be changed, if they be before spiritualized, that they may come into their Sulphur and Argent vive Nature; these are the two Sperms, the one Masculine, the other Feminine complexions, and these are composed of the Elements: the Male Sulphur, is nothing but Fire and Air; and true Sulphur is as a Fire, but not the Vulgar, which is of no Metallick substance; the Feminine Sperm, called Argent vive, is nothing but Earth and Water. These two Sperms, old wise men called two Dragons, or Serpents, the one is winged, the other not; Sulphur not flying the Fire, is without wings; the winged Serpent, is Argnet vive born up by the Wind, therefore in her certain hour she flieth from the Fire, being unconstant in it; but if these two Sperms, separated from themselves, be united again by triumphing Nature in the Book of Mercury, which is the Fire Metalline, then united it is called of Philosophers the flying Dragon, because the Dragon kindled by his Fire, while he flieth, by little and little spreadeth his Fire and poisonous Vapours into the Air: the same thing doth Mercury, which placed upon an exteriour Fire, being in his place in a Vessel, setteth on fire his inside, which is hidden in his profundity; and then may any one see how the external Fire doth inflame the natural of Mercury and shall see a poisonous Vapour to break out into the Air, which shall be of such a stinking and pernicious poison which is nothing else but the

Head of the Dragon, which speedily went out of
Babylon. But other Philosophers having compared
this Mercury with the flying Lyon, because a Lyon
devoureth many Creatures, and recreates himself with
his voracity, these things excepted that resist his
violent fury; so also doth Mercury, which hath in
himself such an operation, that it spoileth a Metal
of his form, and devoureth it: Mercury too much
inflamed, devoureth and hideth Metals in his belly;
but which of them soever it be, it's certain it is
not consumed in his belly, for in their Nature they
are perfect, and more than he indurate: but Mercury
hath in him a substance of perfecting Sol and Lune,
and all imperfect Metals come from Mercury,
therefore the Ancients called it the Mother of
Metals; thence it followeth, where he is formed to
anything, he hath in him a double Metallick
substance.

And first the substance of the interiour, then of
the Sun, which is not like the other Metals; of
these two substances Mercury is formed, which in his
Body is spiritually nourished: so soon therefore as
Nature hath formed Mercury of the two mentioned
Spirits, then it laboureth to make them perfect and
corporeal; but when the Spirits are of growth, and
the two Sperms awakened, then they desire to assume
their own Bodies which done, Mercury the Mother must
dye, which being thus naturally mortified, cannot
quicken itself again as before.

Some arrogant Chymists endeavour in obscure words to
affirm that we ought to transmute perfect and
imperfect Bodies into running Mercury; but a Serpent
lieth in the Herbs: its true, that Mercury may
transmute an imperfect Body, as Lead, or Tin, and

may without labour multiply in a quantity, but
thereby it loseth its own perfection, and may no
more for this reason be Mercury; but if by Art it
might be mortified, that it might no more vivifie
itself, then it would be changed into anything, as
in Cinabar or Sublimate is done; for when it is by
Art coagulated, whether sooner or later it be done,
then his two Bodies assume not a fixed Body, neither
are like to conserve it, as we may see in the pores
of the Earth. But lest anyone should err, there are
in the Veins of Lead some fixed grains of Sol or
Lune, in substance or nourishment: the first
coagulation of Mercury, is the Mine of Lead, and
most fit and commodious it is to bring him unto
perfection and fixation; for the Mine of Lead is not
without a fixed grain of Gold, and which grain
Nature did impart: so in itself it may be
multiplied, whereby it may come to perfection and
plenary virtue, as I have tried and may affirm.

Also so long as it is not separated from his Mine,
that is, from his Mercury, but well kept, for every
Metal that is in his Mine, the same is a Mercury,
then may it multiply itself, so it may have
substance from his Mercury; then will it be like
some green immature Fruit, on a Tree, which the
Blossom being past, is made into Fruit and then the
Apple: but if any should crop away the immature
Fruit, then his first forming would be corrupted,
because man knows not how to give substance or
maturity, as internal Nature, while the Fruit yet
hangs on the Tree, and may have substance and
nourishment from Nature; for so long as maturity is
expected, so long the Fruit draws sap or liquor, and
that by augmentation and nourishment, till it comes
to perfect maturity. So is it with Sol, for if by

Nature a grain be made, and it is reduced to his
Mercury, then also by the same it is daily after an
uncessant manner sustained and reduced into his
place, Mercury as he is in himself; and then must
you expect till he shall obtain some substance from
his Mercury, as it happens in Fruits of Trees: for
as the Mercury of both perfect and imperfect Bodies
is a Tree, so they can have no more nourishment,
otherwise than from their own Mercury: If therefore
thou wouldst gather from Mercury Fruit, which is
shining Sol and Lune, if it be that they be not far
disjoyned, so that it be without long delay, then
think not you as Nature did in the beginning, you
will again conjoyn and multiply, and may without
change augment them.

For if Metals be separated from their Mine, then
they (like the Fruit of Trees too soon gathered)
never come to their perfection; as Nature and
Experience makes it appear, that if a Pear or Apple
be once plucked from the Tree, it would then be a
great folly, if any should again fasten it to the
Tree, and thence expect maturity; for Experience
witnesseth, the more it is handled, the more it
withereth. And so it is with Metals, for if any
would take Vulgar Sol and Lune, and endeavour to
reduce them into Mercury, he would altogether play
the Fool, for no subtle Art is there to be found,
whereby he might not deceive him; although many
waters and Cements, or infinite things of that kind
he should use, he would daily err, and that would
happen to him, that doth them who would tye unripe
Fruits to their Trees. Although some Philosophers
have said well and truly, if Sol and Lune by a right
Mercury be rightly conjoyned, that then they will
make all imperfect Metals perfect; yet in this most

men have failed, who having these three, Vegetables, Animals and Minerals, which in one thing are conjoyned; for they regard not, that Philosophers speak not of Vulgar Sol, Lune, and Mercury, which are all dead, and receive no more substance from Nature, but remain in their own Essence, and can help none other into perfection: they are Fruits plucked off from their Trees before their time, and are therefore of no account, they having nothing more than what they want. Therefore seek the Fruit in the Tree that leadeth you straight unto them, whose Fruit is daily made greater with increase, so long as the Tree holdeth it forth; and this work seen, is great joy; by this means any may transplant this Tree, without gathering his Fruit, and then transport him into moister, better, and more fruitful places, which in one day may give more nourishment to the Fruit, than it received otherwise in an hundred years.

In this therefore it is understood, that Mercury the much commended Tree must be taken, who hath in his power indissolvably Sol and Lune, and then transplant him into another Soyl nearer the Sun, that thence he may gain amicable utility, in which thing Dew doth abundantly suffice; for where he was placed before, he was so weakened by wind and cold, that little Fruit was expected from him, where he long stood and brought forth no Fruit at all.

For indeed the Philosophers have a Garden, where the Sun as well morning as evening remaineth with a most sweet Dew without ceasing, with which it is sprinkled and moistned; whose Earth bringeth forth Trees and Fruits, which from thence are planted; who also receive descent and nourishment from the pleasant Meads. And this is done daily, and there they be both corroborated and quickened, and do not fade; and this more in one year, than in a thousand where the cold infects them.

Take them therefore, and night and day cherish them in a Stillatory upon the Fire; but not with a Wood Fire, or Coal Fire, but in a clear transparent Fire, not unlike the Sun, which is never hotter than is requisite, but should be always alike; for a Vapour is the Dew and the Seed of Metals, which ought not to be altered.

We see Fruits if they be too hot with no Dew, they abide on the boughs without perfection, but if heat and moderate moisture sustain them on their Trees, then they prove elegant and fruitful: for heat and moisture are the Elements of all Earthly things, Animals, Vegetables and Minerals.

Therefore Coal Fires and Wood Fires help not Metals; those are violent Fires, that nourish not as the heat of the Sun doth, which also conserveth all corporal things, because it is natural which they follow.

But a Philosopher doth not what Nature doth, for Nature hath created all Vegetables, Animals and Minerals in their own degree, where Nature reigneth: I will not say that men, after the same sort, by Art make Natural things; when Nature hath finished these things, then by Humane Art they are made more perfect. After this sort old Philosophers, for our information, laboured with Lune, and Mercury her true Mother, of which they made the Mercury of the Philosophers, which in his operation is much more

strong than Natural Mercury; for this is serviceable
only to the simple, perfect, imperfect, cold and hot
Metals; but the Philosophers Stone is useful to the
more than perfect and imperfect Metals. Also that
the Sun may perfect and refresh them, without
diminution, addition or immutation, as they were
created of Nature, so he leaveth them; neither doth
he neglect anything. I will not now say the
Philosophers conjoyn the Tree, for the better
perfecting of their Mercury, as some unskilful of
things and unlearned Chymists do, who take common
Sol and Lune and Mercury, and so ill-favour'dly
handle them, till they pass away into Smoak: and
they endeavour to make the Philosophers Mercury; but
they never attained to that; that is, the first
Matter of the Stone, and the first Minera of the
Stone. If they will come thither, and find any
good, then to the Hill of the seven, where there is
no Plain, they would betake themselves, and from the
highest they have need to look downwards to the
sixt, which they shall see afar off.

In the height of this Mountain, they shall find a
Royal Herb triumphing, which some have called
Mineral, some Vegetable and Saturnal; but let the
Bones be left, and let a pure clean Broth be taken
from, and thus the better part of thy work is done.
And this is the right and subtle Mercury of the
Philosophers, and is to be taken of thee, and first
the white work he will make, and after the red: if
thou have well understood me, both of them are
nothing else, as they call them, but the Practick,
which is so light and so simple, that a Woman
sitting by her Distaff may perfect it; as if she
would in Winter put her Eggs under a Hen and not
wash them, because Eggs are put under a Hen to sit

upon without washing them, and no more labour is required about them, than that they should be every day turned, that the Chickens may be the better and sooner, hatched; to the which enough and more than enough is said. But that I may follow the example, first wash not the Mercury, but take it and with its like (which is Fire) place him in the ashes, which is Straw, and in one Glass, which is the Nest, without any other thing, in a convenient Alimbeck, which is the House, and then thence will come forth a Chicken, which with his Bloud shall free thee from all Diseases, and with his Flesh shall nourish thee, and with his Feathers shall cloath thee, and keep thee warm from cold.

Therefore have I written unto you this present Treatise, that you may search with the greater desire, and walk in the right way; and I have comprehended this small Work in a Summary, that you might the better comprehend the sayings of the Philosophers, which I perswade myself you will better understand hereafter.

FINIS.

CLAVICULA by Raymond Lully

CLAVICULA
OR,
A little Key of Raymond Lullie Majoricane;
Which is also called
APERTORIUM, (the Opener)
In which all that is required in the Work of
ALCHYMY
Is plainly declared.

We have called this our Work Clavicula, or the Little Key, for without this Work none is able to understand what we have wrote in our other Books, in which we have fully declared the whole Art, although with obscure words, by reason of the Ignorant. I have written many and large Books, under diverse Sections and obscure terms, as appeareth in our Testament, where we have handled of the Natrual Principle, where all things are set down that belong to this Art, yet under the Hammer in the proper phrase of Philosophers. Item, in our Chapter in the Philosophers Argent vive, and in the second part of the Testament of the Exuberation of Physical Mines, and in our Book of the First Essence, of the Quintessence of Gold and Silver; afterwards in other Books also made by me, where the whole Art is compleatly set down, but we have hidden the Secret as much as we could. But seeing that no man without this Secret can enter the Mines of the Philosophers, nor make anything that can profit him; therefore by the help of the Almighty, whom it hath pleased to reveal unto me this secret, I will declare this whole Art without any fiction: And therefore see that you do not reveal this Secret unto the wicked,

but unto your entire Friends; though you ought not
to give it to men, being it is the gift of God, who
will give it to whom he pleaseth, and whosoever
shall have it, shall have an everlasting Treasure.
Although Luna receiveth her clearness from Sol, of
these two the whole Mastery dependeth; but seeing
Metals cannot be transmuted (as Avicen witnesseth)
in the Minerals, unless they be reduced into their
first Matter, which is true, viz. that unless you
reduce them into Argent vive; not Vulgar, that is,
not volatile, but fixt, for the Vulgar is volatile,
and full of flegmatick coldness, and therefore it
needeth to be reduced by Argent vive fixed, more hot
and dry, in qualities contrary to Argent vive
Vulgar: Therefore I counsel you, O my Friends,
that you do not work but about Sol and Luna,
reducing them into the first Matter, our Sulphur and
Argent vive: therefore, Son, you are to use this
venerable Matter;   and I swear unto you and promise,
that unless you take the Argent vive of these two,
you go on to the Practick as blind men without eyes
and sence; therefore, Sons, I beseech you walk in
the light, with open eyes, and fall not into the
ditch of Perdition as blind men.

CHAPTER ONE
Of the difference between Argent vive Vulgar, and Argent vive Natural.

We say, that Argent vive Vulgar cannot be the Argent vive of the Philosophers, whatever Art it be prepared with for the Vulgar cannot be detained in the Fire, but by another Argent vive corporeal which is hot and dry, and more digested there: I say, that our Nature is of a more fixt and hotter Nature. than the Vulgar, and that therefore because our Argent vive corporeal is turned into Argent vive current, not teyning the fingers; and when it is mixed with the Vulgar they are joyned, and embrace one another with the bond of Love, so that they never part from one another, as Water mixt with Water, for THUS it pleaseth Nature: But our Argent vive doth enter and mix itself actually with the other Vulgar, drying up its flegmatic humidity, and taking away the coldness from the Body, making it black as a Coal, which afterward it turneth into Powder. Note therefore, that Argent vive cannot shew forth such Operations, as our Physical or Natural, which in all its qualities hath the heat of Nature, and of true temperature, and therefore it turneth the Vulgar into its temperate Nature; nay it doth moreover somewhat else, for after its transmutation, it turneth it into pure Metal, that is, into Sol or Lune, according as it is extended; or from Sol and Lune, as is shewed in the second Chapter or Part of our Practick: Besides this, it hath somewhat greater, for it changeth and converteth Vulgar Mercury into Medicine, which Medicine can transmute the imperfect Metals into perfect: besides it turneth the Vulgar into true Sol and Lune, better than those of the Mine. Mark again, that one ounce

of our Vulgar Natural Mercury, can make an hundred
Marks, and so until infinity, with Argent vive, so
that the Mine shall never fail. Besides this, I
will have you know another thing, that Vulgar
Mercury is not rightly nor perfectly mixed with the
Bodies; for the Spirit cannot be mixed with the
Bodies perfectly, unless they be reduced into the
kind of Nature: And therefore when thou wilt
mingle Lune and Sol in Mercury Vulgar, then these
Bodies must be reduced into the kind of Nature,
which is called Argent vive Vulgar, through the bond
of natural Love, and then the Male is joyned with
the female; for our Argent vive is hot and dry
actually, Argent vive Vulgar is cold and moist
passively, as a Female which is kept in her houses
with temperate heat until the Eclipsis, and then are
made black as Coals, which is the Secret of our true
Dissolution: after they are at last truly knit
together one with another, so that they never part
from one another, and they become a most white
Powder, which are the Males and Females engendered
by true bond of Love; but the Children will
multiply their kinds to infinity, for one ounce of
this Powder, thou shalt make infinite Sol, and
reduce to Lune, better than any Metal of the Mine.

CHAPTER TWO
The Extraction of Mercury
out of the Perfect Body.

℞ Take one ounce of Calx of Luna, let it be
calcined in that manner as is said in the end of the
Work of our Mastery; which Calx or Slime must be
ground into subtile Powder upon a Porphyr, which
Powder ye shall imbibe twice, thrice, or four times
in a day with the best Oyl of Tarter, made in that

manner as shall be said in the end of our Mastery,
drying it in the Sun until the said Calx shall drink
up of the said Oyl, four or five parts more than the
Calx itself was, grinding it always upon the
Porphyrie, as is said: And in the end, let the Calx
be dried up well, that it may well be reduced into
Powder; and when it is well pulverized, let it be
put into a Boults-head with a long neck: put of our
stinking Menstrual made of two parts of Red Vitriol,
and one part of Salt-peter, and let the said
Menstruum first be distilled seven times, and let it
be well rectified, by separating the Earthly Faeces,
in so much that the said Minstrual be altogether
Essential. Afterwards let the Boults-head be well
luted, and put to the Fire of Ashes, with a little
Fire of Coals, until you see the said Matter boyl
and be dissolved: afterwards distill it upon Ashes,
until it loseth the Menstruum, and the Matter be
altogether cold; and when it is cold, let the Vessel
be opened, and the Matter which is cold be put into
another Vessel that is very clean, with its Cap or
Head on, well luted to a Furnace upon Ashes; and
when the lute is well dried, let the Fire be made by
degrees in the beginning, until you get all its
Waters: afterwards augment the Fire until the
Matter be dried, and the stinking Spirits exalted to
the Cap or Head, and in the Receiver: and when you
shall see such a sign, let the Vessel be cooled by
diminishing the Fire: And after the Vessel is
cooled, let the Matter be taken out and made into
subtil Powder upon the Porphyrie, so that the Powder
may be impalpable, which must be set in an Earthen
Vessel well luted and well glazed: afterward put
upon this said Powder common Water boyling, stirring
always the Matter with a clean Stick, until the
Matter become thick as Mustard; and stir the said

Saltish Matter with a Stick, until you see appear grains of Mercury from the Body, and that a great quantity of the said quick Mercury appear, according as you have put in of the perfect Body, that is of Luna; and until you shall have a great quantity, pour upon it boyling Water, and at length stirring it until all the Matter be resolved into a Matter like unto Argent vive Vulgar: let the terrestriety be taken away with cold Water, and dried up by a cloth; afterwards let it give through a Leather, and you shall see wonders.

CHAPTER THREE
Of the Multiplication of our Argent vive.
In the Name of God, Amen.

℞ Of pure Silver three grostes, made into thin Plates, and make Amalgama with four grostes of Argent vive Vulgar, well washed; and when the Amalgama is made, then let it be put into a little Boults-head, with a neck of one foot and a half long. Afterwards ℞ three groste of our Argent vive, formerly extracted and reserved from the Lunary Body, and let it be put upon the Amalgama, made of the body and of Argent vive Vulgar: let the Vessel be luted very well with the best lute, and let it be dried, when this is done, stirring the Vessel exceeding well, that the Amalgama may be well mingled; and thus the Argent vive may be well mixed with the Body. Afterwards put the Vessel in which the Matter is, in a little Furnace, to a little Fire of Coals, and let this not exceed the heat of the Sun, When Sol is in the Sign of Leo, for another heat exceeding that would destroy the Matter, and

the one would fly from the other: and let such a
Fire be continued, until the Matter become black as
coal and thick as pulpis; and let the fire continue
in this degree until the Matter be changed into a
gray brown colour: and when the gray appeareth,
increase the Fire in one point or degree, and let
this second degree continue until the Matter begin
to become white, to the most purest whiteness;
afterwards augment the Fire to the third degree,
continuing it until the matter become whiter than
Snow, and be converted into pure Powder, whiter than
Ashes: and then you have Calx vive, or the quick
Slime of the Philosophers, and its Sulphury Mine,
which the Philosophers have so much hidden.

CHAPTER FOUR
The Property of the said Calx, or Slime.

The said Calx converteth Mercury Vulgar into most
white Powder infinitely, which can be reduced into
true Silver, with some of the Bodies of Luna.

CHAPTER FIVE
Multiplication of the Calx.

℞ The Vessel with the Matter, wherein put two
ounces of Argent vive Vulgar, well washed and dried;
afterwards lute the Vessel well, and put it where it
was before, governing and administering to it the
Fire of the first, second, and third degrees, as
before, until the Matter be reduced into a most
white Powder, and so you may multiply to infinity.

CHAPTER SIX
The Reduction of this Calx viva, into Luna.

When thou thus hast gotten a great quantity of our
Calx viva, or of our Mine, take a Crucible not
covered, in which put one ounce of pure Lune, and
when it is melted, put thereupon four ounces of thy
Powder in small Pills, let thy Pills be the weight
of the fourth part of an ounce: let them be put
upon the middle Luna by degrees, always continuing
the Fire strong, until all the Pills be projected
and melted, together with the Lune, and in the end
make a strong Fire, until it be incorporated:
afterwards project it in an Ingot, and thou shalt
have five ounces of Silver more pure than the
Natural: and thus thou mayst multiply thy
Philosophical Mine as thou pleasest.

CHAPTER SEVEN
Of our great Work to the White, and to the Red.

Reduce the Calx viva, as is said before of Luna,
into Argent vive, which is our Secret. Take
therefore four ounces of our Calx, and reduce them
into Argent vive, as thou didst with Luna, of which
Argent vive thou mayst have at least three ounces:
put this in a little Boult-head with a long neck, as
thou didst before; afterwards make Amalgama with one
ounce of true Sol, with three ounces of Argent vive
Vulgar, and put upon it Argent vive of Lune, moving
it strongly with thy hands, that all may be mingled
together: afterward put the Vessel, well luted as
before, in the Furnace, making the Fire of the
first, second, and third degree: in the first
degree thy Matter will become black, like to a Coal,
which then is called the Eclipsis of Luna and Sol,

and there will be a true commixtion, whereby is
begotten the Sun and Sulphur, which is full of
temperate bloud; after the appearing of his colour,
continue the Fire of the second degree until the
Matter become gray, then continue the third degree
until the Matter appear most white; afterwards
augment the Fire to the fourth degree, continuing so
that the Matter may appear red as Cinnabar, and the
Ashes become red: this Calx you may reduce into the
finest Sol, as is said before of Lune.

FINIS.

SECRETS DISCLOS'D

One Friend to another; as
Bloomfield suppose,
The Philosophers Stone the
Secrets doth disclose.

I shall tell it to you openly: Our Medicine is a
Stone, that is no Stone; and it is one thing in
kind, and not diverse things, of whom all Metals be
made; and so it is no Salts, nor Waters, nor Oyl
combustible, nor mans Hair, nor mans Bloud, nor
Iron, nor Goats-horns, nor Herbs, nor none such
things that discord from Metals, as many Fools
devise: But he is two things, for he is Water and
Earth; not Water of Clouds, nor of Corrosives, nor
Water of Salts, but Water of the Sun and the Moon,
that burns our Earth more than any Fire, And it is
three things, that is, Body, Spirit, and Soul; and
it is four things, Earth and Water, and Air, and
Fire; and therefore he is found in every place , and
in every time. And he is also unstable in colour,
as a shame-fac'd Woman that changeth her colour for
dread of her Love, that reproveth her of untruth;
for now she is pale, now green, now red: so our
Stone is turned to all colours, for he is black, and
white, and pale, and blew, and green, and red; of
this Matter our Medicine is made that we call Ixir,
and Elixir, that is, the Philosophers Stone. Take
this Stone, and put him in a well-closed clear
Vessel, that thou mayst see his working; and when
thou hast Water of Air, and Air of Fire, and Fire of
Earth, then it is done, for the Spirit is departed
from the Body, and leaveth the Body dead and black:

But if the Sepulchre be well closed, he will come in again to the Body and make him rise again to life. and then the Body and the Soul shall ever be together.

And therefore take a Red man, and a White woman, and wed them together, and let them go to Chamber both, and look that the door and the windows be fast sparr'd, for else the Woman will be gone away from her Husband: And if she lye with him right warm on Bed, the beware that she go no where out, for if she do, he shall never overtake her, if he were as swift as a Faulcon; for if she may no where out, she will come to him again, and lye with him on Bed; and then she shall conceive and bear a Son, that shall worship all his Kin, and then will she never after go away from her Husband.

For this Man and this Woman getteth our Stone: But the Man must be fell and quaint to make her to abide with him with meekness, and not with sturdiness; for if he be boisterous to her in the beginning, she will flee away from him, and if he be easie with her in the beginning, she will be his Master a good while. This is a hard marriage, nevertheless one comfort this is, after that she hath born a Child, and known somewhat of disease, she will be the more sober, and never leave him after. But shortly, all our working is no more but take our Stone, and make him rotten in Horsedung, and then seeth him in his own Water, and afterwards fry him in his own Grease, and then roast him till his Grease and his Water be all dried up, and then burn him all to Powder, and then bake him on an Oven till he will melt as Wax, and then thou hast an end. And then thank God that this Work is so easie, for thy Stone is but one

thing, and all one Vessel, and all one working, from
the beginning to the ending: but look that thy Fire
be easie and soft in the Putrefaction, and in the
Solution, and the Distillation, till it be black;
but then strengthen alway till in the Dessication
and the Imbibition, and in the Sublimation, and in
the Coagulation, and the Congelation, and fixing of
the Spirits, and in the Calcination, and in the
Incineration; but in the citrination, and
Rubification, and Inceration, and Liquefaction, is
all their strength. But if thou understand not
this, Friend, meddle thou not of this Art, until
thou have gone better to School; and hold this in
Counsel for my love, as I shall trust to you
hereafter.

        Farewell.

A Philosophical Riddle.

A Strife late rose in Heaven,
    Yet undecided,
And the chief Deities were by pairs
    divided:
Saturn and Luna one Opinion held,
Which Jove and Mercury (combin'd)
    refell'd:
Venus and Mars, that still have loved
    either,
Gainsaid them all, and would assent
    with neither.
In this dire brawl, 'tween these
    three pairs begun,
To Judge and Umpire, they all chose
    the Sun:
Therefore amidst them all, his place is
    still,
With power t' advance and grace
    which part he will,
By all their joynt assents; for as his
    might
Great is, so clearest is of all his
    Light;
And those with whom he holds must
    needs as best
And worthiest, bear the Glory from the
    rest;
And since he needs must joyn with one
    (for odds)
Cannot remain long 'mongst agreeing
    Gods.
Shew me (some man that can) with
    which of these
Three pairs the God consents, and

      best agree
And (on the New Lights word)
      I that before
Knew naught, will rest and ask no
      Question more.

THE ANSWER OF Bernardus Trevisanus,

TO THE
EPISTLE
OF
Thomas of Bononia,
Physician to
King CHARLES the 8th.

Reverend Doctor, and Honoured Sir,

With the tender of all possible Respects and
Services be pleased to understand, that I have
received your very large and copious Letter by Mr.
Awdry, together with the Stone of your most secret
Work; which truly is a remarkable argument of your
Friendship, by which the confidence you put in me
appears manifest and very great, and with how great
and piercing a Wit also you are illustrated. Now
then I shall very willingly Answer unto your
Epistle: Some things I shall approve, which you
have written learnedly and ingeniously, other things
I shall briefly touch, and refute strictly and
Philosophically, but not arrogantly, and throughly
discuss them with submission and respect unto your
Honour, and request: For in this sacred and secret
Art, as in others, the truth of the Theory ought to
be confirmed by Practical experience. Now
therefore, Reverend Doctor, let us visit one another
with such Returns and Treatises, since we may not be
bodily united. But it is your wisdom (as you very
well know) to know and inspect things by their
Causes, for Experience is deceitful when not guided
by a previous understanding. There is necessary to
the Students in Philosophy, a strong and discret
meditation, that the Work they undertake may be

conveniently brought on to its utmost perfection:
For contingent errors happen unto them who will fall
to work, omitting or neglecting the judgment of a
mental practice, which the Theory frameth in the
mind before the operations proceed to the composure
of any Work: For Work must attend Nature, and not
Nature follow Work. He then that would effect
anything, must prepare his mind with the knowledge
of the Natures and eventual Accidents of things, and
afterwards he may safely put his hands to the Work.
And indeed I clearly perceive your mind to be highly
instructed in these things, by your Experiment set
down fully in your Epistle: For as Water which is
cold and moist, if it be well mixt with Vegetables,
assumes another quality, and in decoction takes to
it and puts on it the quality of the thing wherewith
it is throughly mixt; so also Quick-silver assumes
different natures and qualities in things familiar
unto it, and throughly mixt with it: as if it be
joyned to the Sun, the qualities of the Sun; if to
the Moon, those of the Moon; if to Venus, of Venus:
and so in other kinds of Metals. Their kinds
therefore ought to be decocters therein, and Mercury
is their Water, in which by a mutual alteration it
assumes in a convertible manner their mutations.
And this Water contracts unto itself from them a
Nature in a resemblance to Vegetables, dococted in
simple Water: though these kinds are not altered in
their colour outwardly, under the form of fluidity,
in respect of the thickness of the Matter and Earth
immersed in, and united proportionably to the Water
of Mercury; but we find it otherwise in other
diaphanous humidities: For this altered, Nature is
altered, and its colour outwardly is hid under the
appearance of Mercury, and is not manifest to the
sight. And this you at large discuss and show, how

simple River Water is the first Matter and
nourishment of Vegetables, and consequently of all
living and sensitive Creatures: therefore if any of
them all be decocted in it, it assumes and puts on
itself the virtue and propriety of their Nature:
wherefore being in itself cold in the highest
degree, yet by means of things decocted in it, it
works in us the effect of a thing hot in the first
degree, that I may use your words. Moreover, there
is nothing that nourisheth more than the Broth or
decoction of good flesh; and if the Water in which
flesh and Herbs are boyled, or the things boyled in
Water, be eaten moist, or the simple Water after
boyling be taken or drank, it hurts not at all, yea
it will profit and help much, although before in its
simplicity and nature it would have been hurtful.
Now this comes to pass because that Water is not
such as it was before. In like manner Quick-silver
is the Matter of all Metals, and is as it were
Water, (in the Analogy betwixt it, and Vegetables or
Animals) and receives into it the virtue of those
things which in decoction adhere to it, and are
throughly mingled with it; which being most cold,
may yet in a short time be made most hot: and in
the same manner with temperate things may be made
temperate, by a most subtle artificial invention.
And no Metal adheres better to it than Gold, as you
say, and therefore as some think Gold is nothing but
Quick-silver, coagulated by the power of Sulphur,
&c. And thence you would conclude, as I think, and
well, that if Gold be decocted and dissolved rightly
in the natural way of Art, Quick-silver itself will
obtain the natural properties of that Gold. But the
way of this decoction and solution of Metals, is
known to very few, and it manifestly appears: for
the cause of this Solution is the moistness of

Mercury, restrained by the compactness of an
Homogeneal Earth; and contrarywise, the coldness of
the Earth, restrained by a Water Homogeneal to
itself, the Homogeneousness of qualities remaining:
so that there is in it a single dryness, and double
coldness, a simple moistness, but under a
disproportion of immaturity to the anatical
proportion of the ripe digested Sun. The dissolver
therefore differs from the dissolvend in proportion
and digestion, and not in matter: because Nature
might make this of that, without any additional
mixture, as Nature doth wonderfully and simply
produce Gold of Quick-silver, as you have learnedly
discoursed in your Epistle. For in Vegetables, the
moisture of simple Water is taken for an intrinsick
dissolution, that things congealed by Art, might
diffuse into their effects; and the dissolution of
things come about with the coagulation of Water with
the dissolution of things, and contrarywise: and so
it is likewise in the Mineral Water, and things of
its kind. He therefore that knows the Art and
Secret of Dissolution, hath attained the secret
point of Art, which is to mingle throughly the
kinds, and out of Natures to extract Natures, which
are effectually hid in them. How hath he then found
the truth, who destroys the moist nature of Quick-
silver? As those Fools who deform its kind from its
Metallick disposition or dissolution, and by
dissolving its radical moisture, corrupt it, and
disproportion Quick-silver from its first Mineral
quality, which needs nothing but purity and simple
decoction. For example, they who defile it with
Salts, Vitriols, and aluminous things, destroy it,
and change it into some other thing, than is the
nature of Quick-silver: For that Seed which Nature
by its sagacity and clemency composed, they

endeavour to perfect by violating and destroying it, which undoubtedly is destructive to it, as far as concerns the effect of our Work. For the Seed in humane and sensitive things, is formed by Nature, and not by Art, and well mixed; but nothing is to be taken from it, nor added to it, if the same species must be renovated by the procreation of its own kind: so the same Matter must abide and continue, that the same Form may follow, which it doth not otherwise. Wherefore, excellent Doctor, false and vain is all their doctrine, which altereth Mercury, which is the Seed, before the Metallick species be joyned with it: For if it be dryed up, it dissolves not. What then can it do in the solution of things of its own species? For if it be heated beyond its natural digestion, it will not cause nor generate in the Metalline species a Feverish heat as it were, and will impertinently turn cold into hot, and passive into active; and the errour from thence will be incorrigible, and labour lost. For example, Fools draw corrosive Waters out of inferiour Minerals, into which they cast the species of Metals, and corrode them: For they think that they are therefore dissolved with a natural Solution, which Solution truly requires a permanency of the dissolver and dissolved together, that a new species might result from both the Masculine and Feminine Seed: I tell you assuredly, that no Water dissolves any Metallick species by a natural Solution, save that which abides with them in matter and form, and which the Metals themselves being dissolved, can recongeal: which thing happens not in Aquafortis, but rather is a defilement of the Compound, that is, of the Body to be dissolved, Neither is that Water proper for Solutions of Bodies, which abides not with them in their Coagulations; and finally Mercury is of this sort, and not Aquafortis, nor that which Fools imagine to be, a lympid and diaphanous Mercurial Water: For if they divide or obstruct the homogencity of Mercury, how can the first proportion of the Feminine Seed consist and be preserved? Because Mercury cannot receive Congelation with the dissolved Body, neither will the true kind be renovated afterwards in the administration of the Art, nay but some other filthy and unprofitable thing. Yet thus they think they dissolve, mistaking Nature, but dissolve not: For the Aquafortis being abstracted, the Body becometh meltable as before, and that Water abides not with, nor subsists in the Body, as its radical moisture. The Bodies indeed are corroded, but not dissolved; and by how much more they are corroded, they are so much more estranged from a Metallick kind. These solutions therefore are not the foundation of the Art of Transmutation, but the impostures rather of Sophistical Alchymists, who think that this Sacred Art is hid in them. They say indeed, that they make Solutions, but they cannot make perfect Metallick species, because they do not naturally remain under the first proportion or kind, which Mercury the Water allows in Metallick species. For Mercury is corrupted with Metals by way of alteration, not dissipation: because Bodies dissolved therein are never separated from it, as in Aquafortis and other corrosives, but one kind puts on and hides another, retaining it secretly and perfectly: so Sol and Lune dissolved, are secretly retained in it. For their nature is hid in Mercury, even unto its condensation, of which they lying hid are the cause, in as much as they are latent in it: and as Mercury dissolves them, and hides them in its belly, so they also congeal it, and what was hard is made soft,

what was soft, hard; and yet the kind, that is,
Metals and Quick-silver, abide still. He therefore
who thus dissolves, congeals rather, and the
corrupted species conjoyned, receive their old form
by an artificial decoction: Notwithstanding this
dissolution makes several colours appear, because
the species remain as it were dead, yet their
intrinsical proportion is permanent and entire. So
the Lord in the Gospel speaks by way of similitude
of Vegetables, Unless a grain of corn fallen on the
earth do dye, it abides alone; but if it dye, it
brings forth much fruit: Therefore this alterative
corruption hides forms, perfects natures, keeps
proportions, and changes colours from the beginning
to the end: For when the Water begins to cover the
Earth, the black colour begins to be hid under the
white; when the Air covers the Water and the Earth,
the citrine colour appears; which is turned to red,
when the Fire covers the Air, or the other three
Elements. And these last colours abide hiddenly and
intrinsically, and appear under the shew of a white
Spirit in liquid Mercury, until it be recondensed in
the Powder which is in the Bodies: because the Soul
lies hid in the Spirit, as in the condensation the
Spirit and the Soul lie hid in the Powder or Body,
For there is a corruption in the things to be
altered, but no dissipation of parts, unless some
superfluous parts be to be rejected as unprofitable
for the generation, whereupon the Artificer purifies
his Work, that digestion may succeed better. This
is manifest by example in Grain, for of two grains
of Wheat, if the one be cast into good ground, there
it putrifies, dies, and loses its external form, but
nothing thereof is dissipated, yea in its time it
encreases into a multiplicity of Fruit, and there is
indeed made a corruption only of the form, and not

any dissipation of the matter: But if the other grain be cast into the Fire, then both matter and form are corrupted, and the whole is dissipated, and that corruption is unprofitable for generation. Wherefore Water dissolves not Bodies, but those only of its own kind, and by which it may be condensed: nor can Bodies be at all nourished to generation, but by their like, which can preserve the species destroyed by that transmuting Body, through the artifice of the Work: though Vegetables are nourished by things of different kinds, yet before they nourish them, they are assimilated (the dissolution of them being first made) according to the proportion of the things which suck and draw them to them. It must be noted therefore, that the Solution of Metals may be made by different ways: one, which Fools know, as is abovesaid, with Foreign things, which abide not with the dissolved Metals, which is rather to be called a corrosive destruction and defilement of the compound. The second Solution is made by the power and force of Fire, which is no true Solution, but a melting rather of the colligated Elementary parts: for the outward heat of the Fire, in dissolving the Compound, finds out its intrinsical, natural or native Fire within, which internal and proportional Fire dwells in the Air, therefore it dissolves the Air itself. But that dissolved Air resides and dwells in the Water, and the Water in the Earth, and the Water itself dissolves the Earth, so that it melts both the active and passive; but this melting is no Solution, yea it is a dissipation, because the Elements there being homogeneous to one another, and proportionably fixed, by digestion are mixt, and one of them educed out of the power of another generally: And therefore this falls out even in pure Bodies, in

which the Elemental natures are fixed Wherefore in
them the flame of Fire causeth melting, and
dissolves that whole Body to fluidity, and not to a
separation; because Fire cannot flow, unless the Air
consubstantial to it flow; neither doth the Air
flow, unless the Water be dissolved; nor doth the
Water flow, unless the Earth flow: and
contrariwise, as the Earth is dissolved by the
Water, so on the contrary side the Water retaineth
the Air, and congealeth it: and in the same manner
ascending upwards, the Air retaineth the Fire in
Congelation, because the more fixt and fixing
Elements cause fixation, by acting together on one
another; as Earth and Water, and in a contrary
manner Fire and Air, act together each on other unto
Solution. But this Solution is called a melting of
the Compound, and not properly a Solution of it,
because the parts separable from one another in the
generation of the Compound, are not dissolved, as is
done in the third and truly Philosophick Solution,
when the Compound is dissolved in the manner
aforesaid, and yet the parts abide unseparated,
though separable; so that the virtue of the most
digested Elements may be extracted from things to be
dissolved by the dissolver, that is, Quick-silver,
and the grosser parts in such a dissolution acquire
some latitude of subtilty, because the Body is
turned into Spirit, and contrariwise the Spirit into
Body; fixed things are turned into volatiles, and
volatiles to fixed. For this Solution is possible
and natural, that is, by Art of Nature subserving
thereto; and this is sole and necessary Solution in
the Work of the Philosophers which can be done by no
other thing than Quick-silver only, with a prudent
proportion: so as a good Artificer knowing from
within the natures and proportions, ought to make

the proportion from his first entrance upon the
Work. For these two, Sir, are sufficient for this
Work, and nothing else enter it, nor generates and
multiplies as we have said. Besides, you say that
Gold, as most think, is nothing else than Quick-
silver coagulated naturally by the force of Sulphur;
yet so, that nothing of the Sulphur which generated
the Gold, doth remain in the substance of the Gold:
as in an human Embryo, when it is conceived in the
Womb, there remains nothing of the Father's Seed,
according to Aristotle's opinion, but the Seed of
the Man doth only coagualte the menstrual blood of
the Woman: in the same manner you say, that after
Quick-silver is so coagulated, the form of Gold is
perfected in it, by virtue of the Heavenly Bodies,
and especially of the Sun. But by your good leave,
and with respect I must tell you, we must not think
so: For being we are Philosophically perswaded,
that Gold is nothing but Mercury anatized, that is
equally digested in the bowels of a Mineral Earth;
and the Philosophers have signified, that this very
thing is done by the contact of Sulphur coagulating
the Mercury, and by reason of its operation, that
is, from Mercury being digested and thickned by a
proportionate heat. Wherefore we must know, that
Gold is Sulphur and Mercury together, that is, the
coagulant and the coagulated in one: and nothing
added from without thereto, but only a pure
digestion or maturation, which multiplies qualities,
and excites one Element from another out of their
pure possibility into act, no other thing whatsoever
being superadded. But this digestion or maturation
is produced actively, from the superiour Elements,
that is, the Fire and Air, which are not actually
but potentially in Mercury; yet being excited and
assisted by an external heat and by the proper and

natural digesting heat, the passive Elements in Mercury are by them subtilized, being not only potentially existent, but actually, towards Water it self, and the Water is subtilized towards Air, and Air follows to Fire; and in this proportionable action of Nature, and digestion of Mercury, the Male and Female abide together in closed Natures; the Female truly as it were Earth and Water, the Male as Air and Fire: which Earth and Water the Philosophers do mingle in Gold, but called the Air and Fire a Sulphur as it were therein: neither is there any other Foreign addition in the bowels of the Earth. And therefore in Art above ground neither is there found any Foreign addition, to digest or condense Mercury into the nature of Gold, or other species of Metals. Therefore the Philosophers have said, that Sulphur and Mercury make Sol, that is, its corporeity and permanency: And therefore it is not hence concluded, that the external artificial heat, stirring up and assisting the proportional intrinsick heat, to digest and ripen the other two less digested and immature Elements in Mercury, namely its Water and Earth, is of the substance of the Compound. For the external heat is not permanent within, with the quantity and weight of the Matter, nor adds anything thereto: But the intrinsick proportionate natural and simple heat is permanent, with the quantity and weight of the Mercury digested by it; because that heat is an intrinsick and essential part of Mercury itself, to wit, the two more active Elements in it, namely Air and Fire. Therefore Fools do ill and absurdly understand that saying of the Philosophers, that Sulphur and Mercury beget Sol; because, as is sufficiently known, as neither Air nor Fire in the first Mercurial composition, nor afterwards in the natural Metallick digestion, depart nor are severed from Water and Earth, so neither doth Sulphur (which is no other than Air and Fire) depart nor is separated from Mercury, which is the same with Water and Earth. And he is not a natural Philosopher who imagines or asserts the contrary: for the digestion of Gold happens and is made of the first Mercurial proportion, without any addition made thereto by Nature under, or Art above ground, as is said. Neither is that repugnant to what we have said, that a pure Sol and clear Mercury must in this Art be conjoyned, becuase this is not done to that intent to affirm, that there is one Sulphur in Sol, and another in Mercury, or that there is one Mercury in Sol, and another in Mercury, but because the digestion is more mature and perfect in Sol, than Mercury. And also in the Sun the Sulphur is more mature and digested, and therefore more active than in Mercury: whence the Philosophers have affirmed Sol to be nothing else but Quick-silver matured: For in Mercury there are only two actual Elements, to wit, Water and Earth, which are passive; but the Active Elements, Air and Fire, are only potentially therein. But (as it is known) when those Air and Fire in a pure Mercury, are deduced from possibility into act, that is, to a due digestion and proportionable concoction, then it becomes Gold. Wherefore in Gold there are four Elements conjoyned in equal and anatical proportion, in which therefore there is actually a more ripe and active Sulphur, that is, Air and Fire, than in Mercury: Wherefore Gold is by Art dissolved with Mercury, that the unripe may be holpen by the ripe, and so Art decocting, and Nature perfecting the Composition is ripened by the favour of Christ. Whence the cause may be derived, why by the help of the Philosophick

Art, more perfect, noble, and by many degrees more elevated Gold is made, sooner and in less time, than by the work of Nature. Because Nature doth act and work this by boyling and digesting Mercury alone in the bowels of the Earth, without any assistant: which cannot be brought on to the due proportion of Gold, or any other Metal, in a little time. But our Art helps the work of Nature, by mingling with Mercury ripe Gold, in which is a Sulphur excellently digested, and therefore maturing and quickly digesting Mercury itself, to the anatick proportion of Gold, by subtilizing its Elements: whereupon there follows by Art a wonderful abbreviation of this natural Work. Wherefore, my Doctor, I return to the former points; we must not imagine, according to their mistake, who say, that the Male Agent himself approaches the Female in the coagulation, and departs afterwards; because, as is known in every generation, the conception is active and passive: Both the active and passive, that is, all the four Elements, must always abide together, otherwise there would be no mixture, and the hope of generating an off-spring would be extinguished. For in every man, the Masculine Seed to the end of his life is called in him the Agent, when it is first mingled with the Feminine; and whether it be shed out, or consumed in him, Nature for its sake doth vegetate, and is wonderfully increased and nourished, and makes to itself in the same mans loins the like specifick Seed. The like is to be judged of the Feminine Seed in the Woman; wherefore both these Seeds abide always, and are to be esteemed for original Agents, and first Patients. Yet there is a various or different nativity or generation of Mixts and Vegetables: For they are called Simple Mixts, which grow underground, out of

our sight or about the surface thereof, by the
commixture of the Elements alone compounded one with
another: or from their first Solution; because: they
grow not as Vegetables, but how much soever of
matter was compact and mixt in them, so much of
their first weight is reserved in the same
Compounds. For example sake: how much soever at
first a mass of some Mercurial substance doth weigh
in its Mineral disposition in the bowels of the
Earth, so much weight of Gold will abide digested
therefrom: and the Scoriae and Faeces rejected from
it, will rather be diminished than multiplied,
because they receive no nourishment. But there are
manifold degrees of this first and simple natural
mixture: The first is, the naked concretion and
composition of the four Elements, and that
immediate, in which there is not yet any change
made, or exaltation of one Element into another:
but a simple union of a symbolizing composition of
them, presevering and abiding; of which sort Stones
are. The second degree follows upon the first,
because from the aforesaid Stones, Minerals (about
which we discourse) are generated, and the more
noble subterraneous species emerge and arise from
hence: because in these begin the action of
Elements, and their mutual transmutation, though
their action is not in so great vivacity and virtue
as in Vegetables and Sensitives, because they have
neither growth nor sense, as we have said before.
The third degree is that which comprehends precious
Stones and Gems, because in them is found a perfect
and compleat action, from the virtue of the Elements
compacted and acting mutually, as I have declared
more largely in my Philosophy: where I have
perspicuously manifested this third degree, together
with the second, to be a mean betwixt the first and

second composition of Natural things. Then another
nativity or generation is that which is not
accounted to be of Simple Mixts, but Compound
Vegetables: which are truly divisible into four
kinds, or Classes, as I have discoursed more largely
in my other Book which I sent you. For there are
Vegetables, but Sensitives more especially, which
for the most part beget their like, by the Seeds of
the Male and Female for the most part concurring and
commixt by copulation; which work of Nature the
Philosophick Art imitates in the generation of Gold.
No man can artificially perfect any human Seed, but
we can by Art dispose a man to a productive
generation of his like: For the vital Seeds are only
digested in a vegetable manner by Nature, in the
loins of both Parents; but we can by coition mix the
Parents Seeds in natural Vessels, which copulation
is as it were an Art disposing and mingling those
natural Seeds, to the begetting of Man. For example
sake; the Seed of the Man, as more ripe, perfect and
active, is by this artifice joyned with the Seed of
the Woman, more immature and in a sort passive;
which Seed of the Man, because it actually contains
in it the working Elements, to wit, the Air and
Fire, is therefore more ripe and active for
digestion. But the Female Seed doth more actually
contain the undigested and passive Elements, and
which therefore are to be digested, as the Earth and
Water, which being shed out and mingled together in
the natural Vessels of the Female, no Foreign thing
being added thereto, (but the external heat of the
Woman exciting and helping the proportionable inward
heat of the Mans Seed) the active Elements of the
Mans Seed, digest and ripen the Feminine Seed, and
thence a Man is generated compleat and perfect
according to his Nature. So it is in our

Philosophick Art, which is like the procreation of Man; for as in Mercury (of which Gold is by Nature generated in Mineral Vessels) a natural conjunction is made of both the Seeds, Male and Female, so by our artifice, an artificial and like conjunction is made of Agents and Patients. For the active Elements which obtain the name of the Masculine Seed, are naturally conjoyned with the passive Elements, which are as it were the Feminine Seed; but herein the due natural proportion is always to be observed. Now this first Mercurial digestion is called Conjunction, in which the act riseth out of the possibility, that is the Masculine from the Feminine, namely the Air and Fire, from the Earth and Water, by means of a pure digestion and subtilization of them. But the Philosophers and ingenious Artificers imitating Nature, besides this natural digestion of the Seeds in Mercury, have by a most subtle invention made another conjunction and digestion, whence they have not generated simple Gold only, but some other far more noble and perfect thing. For they commanded Gold (in which the Elements are more active) as the Male Seed, to be joyned with Mercury, (in which the passive Elements are existent) that it might be duly dissolved, excluding all Foreign things, save that they used an outward heat, which by helping doth excite the internal natural heat of Gold, to digest actively and ripen Mercury. And so as a Man is generate by Nature, so Gold by Art: Although notwithstanding their Sperm and Seed cannot be generated by Art, because Art knows not proportion of the mixture necessary to procreate Seed; and in Man it knows neither composition, nor mixtion or first proportion, not the causes of subterraneous things, which flow out from the Earth, where is the proper

and natural place of their generation. But those
Seeds produced by Nature are artificially conjoyned,
that out of them in a way of composition, that which
is to be generated may be produced, in which both
the Seeds abide together well mingled, although
Aristotle, as you write, seem to think otherwise.
Wherefore the Masculine Seed of Mercury, or our
Sulphur, goes not away after coagulation, as some
falsly affirm; and that this falls out in Mercury,
by the force of the Sun especially, and that by its
heat chiefly the form of Gold is perfected, as some
think in subterraneous places: Yea rather by the
force of the motion of its Globe, or of its Orb, and
of the whole Heaven universally, because the Solar
Rays do only heat the surface of the Earth, and not
inwardly those its deep places, in which the
generation of several kinds of Metals is brought
about; and neither do the influences of Heaven,
brought down by the Rays, reach into those lowermost
parts, although the subterraneous motion of the
Elements proceed first from the motion of the
Heavens, and not from its Rays of light, nor from
their heat, nor other influences save motion: but
how this comes about, and what is the cause of this
motion of subterraneous things, I believe your
Reverence is not ignorant, and therefore I forbear
it at present. Therefore the Sun is not the
principal cause of Gold, or of its form, though
there be a resemblance in names betwixt them;
because as the Sun is hotter than the rest of the
Planets, so Gold is hotter than any of the Metals,
which the like difference of proprieties. The rest
of the Planets also have obtained like names, whence
this errour of Fools doth arise: For they believe
that every one of the seven Planets, generally and
specially by its influence doth beget one special

kind of Metal, whereunto by a certain propriety it agrees, and is in its nature resembled. But it happens otherwise in subterraneous things, than in Vegetables, in which Heaven or the Sun is the cause of their generation or augmentation, not only by its motion, but also by reason of the heat of its Rays: For the Sun heats the Vegetables themselves, and the superfices of the Earth, the Elements being very strongly reflected by its Rays to the surface of the Earth, because that its Rays can proceed so far. To instance: for that from the twelfth Heaven which obtains the utmost degree of height, proceeding to descend lower, there follow always thicker or less subtle Orbs, till you come to the concave of the Orb of the Moon where alterable things have their place, or the mixt Elements begin, and are terminated under the Hemispheres of things generable and corruptible. And therefore the more subtle and simple Fire is there found, though not altogether pure: because the simple pure Fire cannot be found apart amongst the alterable sorts of things, nor any one of the other Elements, albeit in every Compound thing simple Fire may be found, mixed with other simple Elements, else there would not be many elements, but one only. Therefore the Rays of the Stars of Heaven, of the Sun especially, pass through the foresaid Regions unrefracted, until they descending farther downwards, are reflected in the Fire by reason of its thickness; afterwards descending farther through the Sphere of the Fire, they by moving it reflect the Fire itself into the Air which is thicker. And in like manner the Rays proceeding perpendicularly to lower things, through the Sphere of Air, into the Water thicker than the Air, from which they are reflected back into the Air. And so after its manner they are reflected back by the

Water moved by them, which also is much better
perceived in the Earth, with its thickness above
other Elements. By this decoction and reflection
the Elements are moved invisibly, though not
unperceivably: because we perceive heat by the
motion of the Heavens,and it is always reflected
from the superiour and subtler Elements, into the
inferiour and thicker, unto the surface of the
Earth, by means of the Rays of the Stars descending
perpendicularly from aloft to the lowest things; and
things thus reflected being moved, and by the Rays
of the Sun reflected, accidental heat is produced in
the medium, though sometimes by the Rays of other
Stars, other qualities are produced here below, as
dryness and coldness, as is manifest in Astronomy;
not that the Rays are in themselves hot, but that
they are the cause of heat in such manner as we have
said. Now that these things are true, is manifestly
known from Astronomy and Perspective, whence it is
understood how generations happen in Vegetatives and
Sensitives, thus much therefore may suffice. But
vain Astrologers have other conceits, and think that
the influences of Heaven are from the virtue of its
activity, and not from the virtue of its motion:
which is false, because the Rays of Heaven produce
or effect nothing in the superiour Orbs. For such
Rays cannot be reflected on the aforesaid Orbs, nor
be mixed with them, as they are reflected in the
Elements and mingled with them, not by composition,
but by a moving reflection and mixture of the same
Elements, as hath been said: but in the
supercelestials there is no capacity to receive new
qualities, or Foreign impression, although the Rays
themselves produce wonderful qualities in the
Elements, moved by their reflection. Wherefore, my
Doctor, the Sun in particular is not the cause of

the generation of Gold, nor yet is it by means of
its heat the cause of Vegetables either above the
Earth, or of Mixts about its superfices, which
namely we know to be heated by the Rays of the Sun,
as we have said, which is also agreeable to
Astronomy. But the knowledge of these things, need
not any longer disputation, wherefore I pass on to
what remains; for if you apply your mind to those
things which we have said, you will understand and
you will find it true, that by the activity of
Sulphur digesting and coagulating Mercury, its form
from Gold is specially perfected: but yet you must
not think that from any other Metal, or any Star,
this may be done, as you have written in your
Epistle. That which we have said, is also to be
understood of other Metals, in their kind and
manner; but with difference, because in other Metals
there is a double Sulphur: One which is
superfluous, and may be separated, the form of the
Metal still remaining: Another Sulphur is an
essential part of the Metal, but united to its
Quick-silver, and not separable, so that the form of
the Metal continues: yet that imperfect and
Sulphureous Metal may be perfected by a Medicine
corrupting the form of that Metal, and introducing
another. But what we are to think of the duplicity
of this Sulphur, which you assert in this
Philosophick Art, I pray you, my renowned Doctor,
without violating the Law of our Friendship, or your
Authority, that you would be pleased to consider.
This duplicity of Sulphur is not so distinct in
Mercury coagulated into divers Metals, that one of
them should intrinsically and essentially appertain
to the generation of the Metal, and be esteemed an
essential part thereof, and the other ascribed to
corruption. But there is in every Metallick

species, equally as in Gold and Silver, a simple and single Sulphur; which is termed Quick-silver, from the first Mercurial composition, as hath been declared in the generation of Gold: Becuase Sulphur and Quick-silver are nothing else but the four Elements in Mercury itself, so or so proportionally disposed, as this or that Metallick species requireth. But that which is reputed a second Sulphur, and to be rejected, is a certain Scoria and faeculent part in the Metals, contracted in the coagulation of the Mercury; or a certain superfluity, which being unclean and impure, would not in the digestion of the Mercury, endure a congelation to the form of a Metal: because it was not of an homogeneal and proportionable Nature of Mercury, apt to be congealed and digested into a Metal. But some Philosophers have called this Scoria, a combustible Sulphur, because it cannot subsist, but vanisheth in the testing of Metals, or is separated from them into Faeces. And here I may bring this example: the bloud in Sensitives, and sap in Vegetables, in their coagulation have several and different offices; because some parts of the bloud have a conformity unto Flesh, and therefore may be coagulated and turned into Flesh, and retain the uniform nature of Flesh, and obtain the name of Flesh. But some parts thereof residing in the pores, are of a superfluous humour, which can in no wise be converted into solid Flesh, and therefore are ejected by Sweat and Medicines, and separated from the true Flesh. But in Sanguine complexion there are many fewer superfluities, than in others: So we may conclude by way of resemblance, that it is in Gold and other kinds of Metals; that the purer or impurer Mercury, in its first coagulation, contained or contracted more or less superfluities, or natural impurities. Wherefore the difference is made in the coagulation of Mercury, which specifies and causes divers Metals; and whatever Mercury there is in any sort of Metal, is termed incombustible, and inseparably permanent, though in fixed Bodies it is made volatile by Art, yet by Nature it remains inseparable in an Elemental proportion. But what dross soever was contracted in the Mercury, and mixed with it from the beginning, (that is, in the congelation of Mercury in its first composition, by heat digesting it to a Metallick kind; and therefore it is by the rest taken away from the Mercury, that is the homogeneous Mercurial nature, and separated from the Metallick kind as rejectaneous and heterogeneal) this is not properly called a Sulphur, but a dross and certain superfluity: because Sulphur is nothing else but a pure act of Air and Fire, warming and digesting, or decocting the Earth and Water in Mercury, proportionable and homogeneous unto it. But the dross is that which in the first composition was not pertinent unto the nature of Mercury, nor had a proportion to any Metallick kind in the composition and digestion of the first Elements in Mercury. From these things it is known, that there are not in other sorts of Metals any distinct or more Sulphurs, than are in Gold and Silver, but one only and simple Sulphur; though there are in them more and greater superfluities, than are in Gold. From hence the truth of your saying is known that Gold, of all Metals cleaves most unto Mercury. Now this comes to pass by reason of the purity of both, because in them is less dross, dregs, or superfluity, than in others: For everything doth naturally desire, by a through mixture and union, to be joyned to a thing of like nature to it, and proportionable in homogeneity, rather than with a thing unequal and unlike to it, as we know; like as Water very easily and without contradiction is quickly joyned to another Water, with an identative and uniting mixture. Now in Gold there is nothing but Mercury, therefore being there is in it little dross, (which is not of a Mercurial nature, as we have shewed) there is therein no great resistance, but that a pure Mercury may more easily adhere to Gold and Silver, than to other Metals, in which many superfluities and dross do forbid and hinder other Metals, or their congealed Mercury, any contact, or through mingling with crude Mercury. For those superfluities, as we have already said, are not of the first composition of Mercury, nor of the same natural or proportional homogeneity: and if happily they be of its composition, yet they are not of its proportion; for whatever is of any things proportion, is not superfluous. Wherefore they cannot be inseparable throughly mingled, neither with Mercury to be coagulated by Art, nor with Mercury coagulated, which in the nature of its Mineralness is joyned with them in the same kind of Metal; being such dross is combustible by Fire, and therefore separable. What wonder is it then if in those Metals to which they are accidentally superadded, they hinder their natural commixtion, and permanent union which coagulated Mercury, or other crude Mercury? For this very cause Gold itself, though never so pure, can far more difficultly abide with, be joyned and adhere to an unclean and drossie Mercury, coagulated or not coagulated, than with a pure and clean one. Because a simple Nature doth rejoyce in the society of, and is perfected by a simple Nature, that is like to it, and same with it in its first homogeneity and Elemental proportion: but Gold, as hath been said, is nothing else but Mercury thickned by its proper digestion, and Elemental action: therefore albeit in the Earth there be a difference betwixt Gold and Mercury in ripeness, (because Gold is more ripe than Mercury) yet there is no diversity in their Matter. Therefore whatsoever Gold hath acquired by the digestion it hath unto maturity, Mercury may acquire the same without any extraneous thing. But Art to breviate and contract the Work, joyns Gold with Mercury, as is said, and out of two Sperms it makes and generates artificially that same thing, which Nature doth create in the Mines of one actual Seed, the identity of the Matter being always everywhere observed, but not the same active power. And therefore as nothing extraneous to its Nature, doth enter this Work in its first composition, so neither doth anything multiply it, which is not of the first temperament thereof. Wherefore some men think falsly, that the Philosophers Stone may be composed of divers things, or of all things, and be nourished by them, instead of the aforesaid Sperms, notwithstanding divers names have been imposed on them. Neither doth this Philosophick Work eat anything, or convert it into its own Nature, which is extraneous, because it doth not vegetate. Wherefore though there be in the said Philosophick Stone, a Body and a Soul, or a Spirit, it is not therefore vegetably animated as Trees and Plants: For this Stone, as all Minerals, is of the aforesaid first, and not of the second, or any superiour intention or imposition. But Trees and Plants are of the second imposition, as Vegetables are of the third, fourth, fifth, or last imposition, for mixt things in those four last impositions, do vegetate. For in them the Elements by many transmutations, and by being oftner alterated, are more subtle;

wherefore they are more active and perfect, though they are not more durable and permanent in their permixtion, because the Elements in them are not of a fixt, but dissolvable composition; wherefore they take in their nourishment vegetably. But our Stone, as also all the Minerals, is of the first imposition; because it vegetates not, nor is vegetably nourished, but nourishment befalls it rather by apposition of a nourishment of a like nature to it, and not by vegetation. For example sake: because, as is manifest by experience, out of a Feminine Seed, to wit out of Mercury put to it unitively, insensibly and by way of composition this Philosophers Stone is nourished, by means of a digestive heat. For it takes and assimilates its like unto itself, to be multiplied by way of apposition, and not vegetably; wherefore it becomes weightier in quantity, and more active and perfect in quality: neither doth Fire or heat multiply this our Stone, as its due nourishment, because it is not of its first composition, but heats it by an extrinsical accident: For how can Flame or Fire multiply the Stone itself, or make it of itself more weighty, when it cannot be fixedly and permanently mingled with it, nor is not of its first composition or form? Nothing therefore nourishes and multiplies the said Stone, to the generation of the same form, except the Feminine Seed, which nourisheth it by means of heat, and nourishes it not vegetably, but by way of apposition and commixtion. He therefore who thus multiplies and nourisheth it, shall not erre, because this multiplier and nourisher is turned into the same kind. A man may indeed increase the Stone and its weight by extraneous things; but this must be done out of its natural kind, not convertible into it: For that weight would be made besides Nature, that is, not into the same species, nor into the unity of one species, yea it would be an aggregation of divers kinds, and an accidental composition, which might be separated by the Test. But when the Philosophers said, that the Stone might be made of everything, truly they understood it not, (as some perversly interpret them) that the Stone might be made of divers things, unlike unto it both in kind and nature; or, which is more absurd, that it might be multiplied by a Flame ministred to it from without: for this reason especially, because Fire and its Flame may by a certain production arise out of everything: Now the refutation of this opinion is manifest from what hath been said before. But when the Philosophers say, that the Stone is made of everything, they mean, that it is made of the four Elements proportionally equalized to one another by a due and natural digestion: out of which four Elements everything that is generable and corruptible is made. Therefore by this similitude the Philosophers say our Stone is made out of everything, that is, out of every Element; because if any one of them were mortified or destroyed, the whole proportion of the Golden Nature would perish, and its kind: and everything in whatsoever latitude and sort of alterables, is generated out of the four Elements either actually, or potentially mixt: yet it cannot be properly said of every producible thing, but of our Golden Stone, and other things equally mixt, that they are made out of everything: for this reason especially, because in those things which are not produced by an equal, but by an adequate proportion of the Elements, all the Elements are not actually existent; but in their adequate activity and passion: for some of the Elements are therein either in an active or passive power, and the rest are therein actually. But in the Philosophers Stone, which is Gold, being it is an uniform Work of Nature, all the four Elements active and passive are actually therein, and permanent in an equal proportion. For the Essence or Nature of Gold, is nothing else but the four Elements equally mixd; not that their form and matter may be said to be therein equal, but their passive and active power; that is, they are each alike and equal not in quantity, but in quality: because that the active doth not exceed the passive in its acting; nor on the other side, the passive doth not exceed the active by suffering more: because there is an equal proportion as to measure in our Gold, or in our Medicine, double hot, double moist, double cold, double dry, and all these are actually therein, by actual action and passion; that is, Fire, Air, Water, and Earth, as we have said before. And all these are said to be alike, and equal in quality, not quantity, because they are equal in actives and passives; and they are therefore durably permanent in Gold, because the passive in it consists permanently in its active, and on the other part the passive rises not up against the active. And they ought not to be alike in quantity; that is, there ought not be so much matter of Fire, as there is matter of Earth: because then the Fire by reason of its quality, would be everywhere of an unequal activity with its passive Earth, and of a far greater. Wherefore there is in Gold, as to its matter, but not as to its quality, much more of the heavier and more passive Element, that of the lighter and more active; that is, more in quantity: there is in it a greater quantity of Earth, than Water; a greater quantity of Water, than Air; a greater of Air, than Fire:

wherefore it is the heaviest of all Metals. But in this unequal proportion of quantity there is an equal and like proportion of quality, of hot, dry, moist, and cold, because each of these is in Gold, as hath been said. The cause of which weight is the permanency of the solidity of the Earth and Water, and the solution of an homogeneous water with the Earth because Water dissolves in Homogeneal Earth. Also their intrinsical thorow mixture in their very least particles, is the cause of the weight; because the Water as well in Gold, as Quick-silver, suffers not the Earth to have any pores in it: which is otherwise in other Metals, in which pores are insensibly made in their congelation, because of the dross mingled in those Metals all over, rejected by the Mercurial nature and heterogeneous: whereupon their lightness results, which is nothing else but want of matter, and porousness of the same, as weight is nothing else but a solid addition of matter. Wherefore if there were in an equal commensurative quantity, so much of the solid matter of Fire; as there is of the matter of Earth, Fire would be as weighty as Earth. But the cause of the weight of Saturn, is its immature congelation, because it doth not yet reject the dross of its parts, whence the pores are made in it; but the pure and impure abide through mixt together in it everywhere, as in the first crude Quick-silver, in which the inspissation and coagulation is weak, for that cause Saturn or Lead retains the weight of its Quick-silver, not because of the purity of its solid matter, but because of its immature coagulation or coction. Wherefore if in this Work you would not destroy the Fire and Air, you must preserve in a distinct and like proportion the heat of the Compound: But if you would not destroy the Air and

134

the Water, then in the same Compound you must cherish the humid: so in the same manner you may preserve the Water and Earth, or the Earth and the Fire, in the said Work, by preserving rightly, and by the artifice of the Philosophick skill, both the cold and dry: because if you destroy any one of them, the proportionable form and kind of Gold is lost. For this cause the Philosophers say, our Gold is made of everything, that is, of every Element, every Element being intrinsically preserved in it, and actually compounding it: wherefore all the Elements are intrinsically in act or power, the principles of all compounded alterable things, and for that cause are said to be all things. Furthermore, my Reverend Doctor, for your credits sake, you must understand the sayings of the Philosophers according to the possibility of Nature, and not according to the sound of Words: For they have handled this holy and hidden Art, and its Secrets, under Similitudes, Fables, Riddles, and obscure words, and have hid it purposely, that it might not be exposed to the unlearned, impious, and unworthy. Furthermore, that I may go on to other Heads of your Epistle, I understand the artifice of your Stone to be a composure from Gold, but from your writing I cannot apprehend it, because you set not down the first original of that Composition. Therefore I shall not need to handle it more at large, till you instruct me fully and more plainly in its Composition and Operation: For I cannot neither believe that the Elixir, or Philosophers Stone, can consist of the signs appearing in it, and of the properties of the nutritive vegetation of the flaming Fire, which you attribute to it, as I have openly shewed in what I have said already. But when I received your Work, and the gift of so great a

Secret sent unto me, I at once understood your
unfeigned love, and free confidence in me.
Wherefore for your Friendship sake, I reserve your
Stone with me, and keep it as a most acceptable
gift, and shall write unto you more concerning it,
when you shall declare it to me more manifestly.
But whereas you say, that in your Stone there are
three, a Body, Spirit, and Soul, (which is manifest
to you by your experience and work) the Philosophers
when they said those three natural things were in
their artificial Stone, understood it by way of
resemblance and experiment: For they called the
Earth, its Body and Bones; because it is an
astringent Compound, and restrains the fluid
Elements from their raw flexibility, having the Fire
also with it symbolically by its driness, But they
called the Water and Air its Spirit; because they
are the Elements that moisten and dissolve the
Earth. But they called the Air and Fire, the Soul;
because they ripen and digest the whole Compound.
And they named them thus, with resemblance unto
Human nature, because in the well-constituted Flesh
there ought to be Bones to sustain the Body, and
likewise there ought to be in the Flesh a vivacity
of vegetable Accidents, which are called its
Spirits: contrary to the errors of the Pagan
Philosophers, who thought the vital Spirits to be
something distinct from the Body compounded, and
parts compounding it: so also there must be in
Humane Flesh an informing Soul, digesting in man the
brutal acts, and to work in him the intellectual
work. But we must understand it otherwise in our
Stone, in which the Earth hath the name of the Body,
Air and Water obtain the name of Spirit, neither is
in it a Soul but because it contains the Air and
Fire; which I perceive well, you do perfectly

understand. But the Philosophers divided them in this manner: By a crude Spirit, they extracted a digested Spirit out of the dissolved Body, and they had remaining a fixed mass of Ashes to be farther dissolved, in which they found an incombustible and stoney oyliness and gumminess, which they called the Soul; which enlivens, unites incerates and produces united Natures; and in the Spirit they disjoyned the Natures, so in the Oyl they re-conjoyned them. For our Stone hath not an informing nature, as a Vegetative or a Sensitive, but it hath only a formed form, which form is the very Elements themselves, because it is homogeneous. But mans Body, and that of other Sensitives, is heterogeneous: For Bones, Flesh, Bloud, Marrow, Hair and Nails, are distinguished differently in it; which is otherwise in Gold, in which whatsoever there is, is found to be of one kind. Wherefore, my Reverend Doctor, the Philosophers speak this by way of similitude, by reason of the administration of Art, and operation of Nature: not because there is a Soul in the Stone, but metaphorically, (as you well know) nor Spirit, nor Body, (as an informing form) as it is found in Man, and other Sensitives. Verily I tell you, that Oyl which naturally incerates and unites Natures, and naturally induces the Medicine into other Bodies that are to be tinged, is not compounded of any other extraneous thing, but out of the bowels of the Body that is to be dissolved: which Oyl retains the colour of its Spirit always, until it be rethickened, and then first of all it puts on the Royal Ensigns, that is a citrineness and Metalline form, which it manifests to all; in Gold, a Golden, in Silver, a Silver colour and form: which Oyl if it be Sol, being dissolved, is perceived to be red inwardly, though outwardly it appear white, under

the form of liquid Quick-silver. Now some think to
compound an Oyl as generous and powerful as this Oyl
is, namely out of Mercury throughly dryed, or out of
the substance of Tin, or Body of the Sun, commixed
with ingredients of divers kinds; but for what
concerns our Work, their Experiment is fallacious.
They can indeed reduce the species of Metals into a
kind of Oyl, but they cannot at any hand reduce them
into a Metallick kind, observing and keeping the
proportion of the things to be mixed sound and
entire. But that Oyl may be profitable for Medicine
to sensitive Creatures, because the nature of Gold
is dissolved therein; but yet impertinently and
unprofitably as to our Philosophick Work. Besides,
my Honoured Doctor, that I may lightly touch on
the remaining Heads of your Epistle, you must
diligently and wisely observe, that Fire and Azor,
wash Laton: But Azor is not raw Quick-silver simply
extracted out of the Mine, but it is that which is
extracted by Quick-silver itself, out of the
dissolved Bodies; which is found to be more ripe
upon tryal. Wherefore if Laton be an unclean Body,
it is depurated by such an Azor, which you write
that you have had formerly; and by this Laton
purified by Azor, we make our Medicine for curing
every sick person.   Indeed this Azor is made of the
Elixir, because Elixir is nothing else but a Body
resolved into a Mercurial Water; after which
resolution, Azor is extracted out of it, that is, an
animated Spirit. And it is called Elixir, from E,
which is out of, and Lixis, which is Water, because
all things are made out of this Water: and Elixir
is the second part of the Philosophick Work, as
Rebis is the first in the same Work. But the
Tincture constitutes the third Work; for as the
matter of this Composition produces divers effects,

so it obtains different names one after another. Thence it manifestly appears, that Azor is not requisite to the Elixir, because in this Work the Elixir goes before Azor, and not the contrary; like as Water precedes the Oyl, and the Spirit the Soul: For Azor is drawn and extracted our of the Elixir, as Oyl out of Water, and not contrariwise; as mention is made elsewhere. For example sake; as in the Art of Physick, pure simple Fountain-water, by boyling in the first concoction, is joyned with the Flesh of a Chicken, and thence in the first degree of concoction we obtain a Broth, a good and perfect decoction, and the humid, watry and airy parts of the Chicken being actually dissolved in the aforesaid Water; though there be other Elements therein also actually. But that it may be made a much more perfect Medicine, and more generous for restoring man's sick Body unto health, the decocted Body of the Chick is beaten into a mash, with the said Water already altered into a boyled Broth, or with part of it, and is distilled by a stronger decoction, whence a Broth and decoction will be made much more noble and generous, partaking of the whole nature of the Chicken: Because by this second decoction not only the moist parts, but the hot parts, that is, its aerial and fiery parts, being melted into the Broth or decoction, are throughly mingled and dissolved: and therefore the whole virtue of the Chick is in such a decoction extracted into the aforesaid Liquor. So it falls out in the Philosophick Work, because the crude Mineral Spirit, like Water, is joyned with its Body, to dissolve it in its first decoction: whence it is called Rebis, because it is compounded of two, or a double thing, to wit, of the Masculine and Feminine Seed, that is,

of the thing to be dissolved, though it be one thing
and matter: whence the Verses,

> Rebis is two things joyn'd, yet it's
>   but one
> Dissolv'd to their first Seeds, the
>   Sun or Moon.

Now out of these two things dissolved together, the
Elixir is compounded, that is, a tinged Water:
whence the Verses,

> Pure Bodies are of Lixis Made by
>   Art;
> Hence Greeks Elixir term its second
>   part.

Out of this Elixir, my Venerable Doctor, as out of
the first Broth or Bullion of a simple decoction,
Azor is extracted, to wit, by a stronger and
iterated distillation: which Azor resembles and
participates the nature of its Body from which it
was extracted, which is hot, and retains its virtue
in itself, namely an Oylie nature, which is hot and
moist, because it is actual Fire and Air; though all
the Elements are in it in Essence, and by
Composition. Medicines therefore to cure the Bodies
of Sensitives, may be composed out of the said
Metals by several artifices; but they are not
pertinent to the Philosophick Work, as the Elixir is
to Azor: that is, the vital Spirit and fugitive
Soul are not diaphanous, nor transparent as the
clear tear from the Eye: nor every dissolving
Spirit, though they be each of higher Natures than
another, according to their degrees, as the Soul is
higher than the crude Spirit, being they are not of

one form. For as the Soul lies hid under the
species of a dissolved Spirit, before its re-
inspissation, (for the Soul being extracted out of
the Body, always appeareth like Quick-silver) so
after its inspissation the Soul and Body lie hid
under the species of a Body. Your Worship hath seen
an Experiment thereof, in the powder sometime sent
to that King whose Physician you are; in which
Experiment, Quick-silver was found in the species of
Quick-silver, but if that which remained in the
bottom had been coagulated, it would certainly have
assumed the same form of Powder: But that Powder
must be called A Tincture nominally only, not that
it is a Medicine for Metals, for it is not yet
perfectly fixt; yet as a Medicine for Men, it is of
very-good force. But the fixt Medicine without all
doubt exceeds this humane Medicine in all virtues,
both as to Metals, and to Men; which cannot come to
pass in a clear diaphanous and transparent Liquor:
Because if the aforesaid Elixir and Azor, that is,
Spirit and Soul, did appear in, and had a
transparency, now the Earth as to its proportion had
left the Water, and had been separated from it,
which had thickned and coagulated its parts, causing
an opacity in the Elixir and Azor, and making a
congealable Metallick form to consist. For in the
condensing of fixed Metallick species, the condenser
must act upon the condensable, and the coagulating
upon the coagulable; which cannot be in the
aforesaid diaphanous and clear Water. But it
happens otherwise in Vegetables, in which a simple
and diaphanous Water is thickned by decoction into
the Vegetables themselves: which yet by the Test of
the Fire doth at length vanish and evaporate,
because it is not permanent and fixed in its
composition, because it had not with it an Earth

Naturally homogeneal to it in its composition, as
Quick-silver hath: which Earth indeed is the cause
of permanent fixation in homogeneous things:
wherefore simple Water cannot by coagulation be so
fixed with Vegetables, as Mercury with Metals. If
therefore Mercury should be reduced to a
transparency in the Work of the Philosophers, it
would by good reason remain of an uncoagulable
substance; nor would it be congealed upon Laton to a
Metallick form, species and proportion, which
carries not with, nor in itself its own congelation,
namely Water the Earth: which Earth (as we said) is
Mercurial, and the first cause of Inspissation,
Coagulation, and Fixation. If then this Water abide
distitute of Metallick proportion, how should it be
possible that such like species should be produced
from this Composition? They also erre who think to
extract a limpid transparent Water out of Mercury,
and out of it to work many wonderful things: For be
it so that they can perfect such a Water, that Work
would conduce nothing either to Nature or
proportion, nor could it restore or build up any
perfect kind of Metal: For so soon as Mercury is
throughly changed from his first Nature, so soon he
is forbidden entrance into our Philosophick Work,
because he hath lost his Spermatick and Metallick
Nature. From these things it is manifest, what
truth there is in your opinion, and in what it is
contrary and improper, when you say, there must be
had (as I think) to perfect the highest Elixir, a
Gum in which are all things necessary thereunto, and
containing the four Elements, and it is a most clear
Water as a tear from the Eye, made Spiritual, &c.
which make Gold to be a mere Spirit: For a Body
penetrates not a Body, but a subtle congealed
Spiritual substance, which penetrates and colours a

Body. Let it be so as you say, my Venerable Doctor, that Natures are not joyned but in a Gum or Oylie substance, and equal proportioned, having a Spiritual Nature, the Elements being yet fixedly shut up in it; unto which Gumminess the whole Philosophers Stone is at last reduced by Inceration, under a gentle flux, after the manner of an Inceration resembling all the Elements, standing like Copper and in the nature of Copper, existing also in a subtle Spiritual Nature penetrating and colouring Metallick Bodies. For this Stone in the sublimation of the first crude Body, hath not lost its kind, namely of the same Spirit, neither yet in the perfect and great Gum doth it lose its first Nature: Therefore Gum and Oyl belong not otherwise unto this Work, but as Elements equally proportioned shut up together, resolvable, united in the Oylie viscosity of the Earth, retained, burned, inseparably mixt. For this Gum of Oyl first is extracted out of the Body, drawn into an incinerated Spirit, till the superfluous humidity of the Water, be turned into Air, and one Element be excited from another Element by digestion, and what was of an Aqueous form, become of an Oylie nature: and so the whole Stone at last assumes the name of Gum and Sulphur. For Geber teacheth this, when he saith, as you have written in your Epistle, if any person know to joyn and friendly unite our Sulphur unto Bodies, he hath found one of the greatest Secrets, and one way of perfection: as if he should say, If any man can reduce a Body to this, that it may be made a Gum which may be throughly mingled with other imperfect Bodies, he hath found the greatest Secret of Nature, &c. because this perfect Stone is a Gum and a Sulphur, as is known by what we have already said. But you must know, that Geber with highest prudence and wonderful artifice hides the truth under a Veil, intermingling with it many obscurities and falsities, which those who are ignorant at first appearance imagine to be truth: yet he speaking like a Philosopher secretly under his craft, doth openly, learnedly and Philosophically describe the truth: wherefore the unexperienced and Sophisters, not understanding his mind and wit, nor the nature of the thing, do perversly turn aside to the vulgar exposition and sound of the words. For he saith, If thou knowest that, we have said something to thee; but if thou knowest not, we have said nothing to thee. Wherefore in reading Philosophick Books, consider especially the possibility of Nature; notwithstanding some writers of this Art have also sometimes erred, and have happened sometimes to have handled it, as to the natural truth either ill or ambiguously. As it may be observed that Arnoldus de Villa Nova hath said, in a Book which he called his Rosary, that raw Mercury, that is, Quick-silver, which in its own nature is cold and moist, by Sublimation may be made hot and dry; afterwards being revived, it becomes hot and moist like the complexion of Man. You will say then, what wonder it is if it be joyned with the Sun, that it likewise becomes of the nature of the Sun? For Mercury is of a convertible nature, as the Heavenly Mercury, which is such as the Planet is with which it is in Conjunction. For that Arnoldus, though in other Sciences he were a Reverend and Ingenious Doctor, yet in this Art he handled Experiments only, without the learning of the Causes. Now when he saith, that in the first Sublimation the crude Spirit is sublimed from the inferiour salt Minerals, and that Mercury itself, which in its own nature is cold and moist, becomes a Powder of a hot and dry nature, as

he saith, this yet conduces nothing to our Work.
But let it be so, that he makes of Mercury such a
Powder as he speaks of, that is, thoroughly dried
and hot by sublimation from Salts; yet these
Purifications are vain and impertinent to our Work,
yea as to the perfecting of our Work they are
hurtful. For though these inferiour Minerals
communicate with Metals in their nature, yet not in
kind and proportion: For the superiour and
inferiour Minerals, in their nativity and
subterraneous formation, are of one and the same
constitution universally, and therefore of the same
nature; but they differ in proportion, quality, and
kind or form. Wherefore if Mercury be distilled
with those inferiour Minerals, and throughly dried,
then his internal nature is confounded and
disproportioned, and is hindred and made
unprofitable, as to the effect of a Feminine Seed,
and invalid for our Metallick Work. For so soon as
he is turned into the form of a Powder, (except from
his Body of Sol or Luna) so soon he undergoes a
thorough driness, unprofitable to the Philosophick
Work. Yet I deny not, but that a drosse and impure
Mercury may and ought, by a simple Salt, be sublimed
or purged once or oftner, according to a due
Philosophick experience, to take from it its dross
and outward Mineral impurity, so that
notwithstanding the fluidity and radical humidity of
Mercury may always remain unaltered: For the
Mercurial kind and form in such a Work, ought to
remain uncorrupted, as hath been said already. Nor
ought its outward from to be reduced into a
thoroughly dried Powder; because its external form
being corrupted, shews its internal nature to be
confounded, unless it be in the way of generation
that it be altered, as may be manifestly seen in the

signs which appear in the Work of the natural way. For there are Sublimations of Mercury from its own proper Bodies, which are conjoyned and mingled with it, by an Amalgamation which it in its most inward parts, from which being oftentimes raised and reunited, it rejects and loses its superfluities, and is not confounded in its nature; and afterwards it is very agreeable to the Philosophick Work, and powerful to dissolve Metallick species; yet it is not greatly altered intrinsically for the Philosophick Work, unless it be altered by fixed Bodies dissolved in it. But wonderful things may be done in Medicines for Sensitives from this dried Powder, whether it be reduced into an Oyl, or into Water, or it abide in a Powder; but it is not at all pertinent to the Philosophick Experiment. And therefore it must be universally noted, that so soon as Mercury is turned into a Powder, of whatever sort, contrary to the nature of its Body to be dissolved, so soon will it be unprofitable to the Philosophick Work. There are certain deceiving Sophisters, who by joyning Venus to it, or adding other species, make a Sophistick Work; that is, they give unto imperfect Copper a colour, but not natural; they induce indeed a kind of an apparency, but not a true nature, that is, transmutation: like as he that paints a dead Image, or composes a Statue of Wood, which appears only, but is not; and as much as a living differs from an Image and Picture, so much differs their Work from the Philosophick. Hence this mixture perseveres not in the Test of the Fire, though it be Mineral; because Nature attracts it not from a proportionable digestion, nor hath Art vehemently decocted it to an alteration of the mixt natures: wherefore that Copper appears to be superficially only, and not permanently and

intrinsically tinged. Wherefore we must not adhere
to the Experiments of deceitful Sophisters, because
the truth of the natural Art confutes this
Sophistick Work, and shews it to be false. And if
you will instance farther, and say, that as the said
Arnoldus by Sublimation purged away the dross of
Mercury, and dried it in its nature; so also (as
you say) he by reviving it, moistned it again, and
made the Mercury itself hot and moist, and in its
nature conformable to its Body. This hinders not
(my Reverend Doctor) nor refutes the truth of the
Philosophick Art, yea rather an errour appears in
the Natural Art: For, as is manifest, Arnoldus doth
teach, if you regard the sound of his words, that
Mercury thus throughly dried, is revived by hot
water into which it is cast; and he saith that it is
made hot and moist, when it was first sublimed hot
and dry. But what true Philosopher would say, that
Mercury or any other Metal, is changed in nature and
internal quality by simple Water, however hot or
boyling, or that it could thence acquire its natural
humidity, and so be revived? Therefore Mercury in
this revival acquires nothing, because common water
neither decocts nor alters it, because it neither
hath entrance nor ingress into it, and that which
neither hath entrance nor ingress, alters not;
because everything to be altered, must first be
throughly mingled. For indeed such a Water may wipe
away from it some superficial dross swimming upon
it, but cannot infuse into it a new quality: For
what nature soever Mercury reduced into a Powder,
and mortified by Sublimations, retained, such nature
altogether it retains revived by Water. Now this I
would have to be spoken in honour and respect unto
the said Arnoldus; but I contemplate and defend the
truth of Nature and Experience. Furthermore,

honoured Doctor, that I may by this my Answer
satisfie your Epistle, and put an end thereto, I
humbly entreat you that you would take in good part,
and favourably bear what I have written, not by way
of Confutation, but Disputation: But if I have
answered any thing that offends you, take it yet in
good part and favourably, or signifie it to me in
writing, and I will satisfie you to my power as the
most true Doctor our Lord Jesus Christ, the Son of
God, blessed for ever and ever, shall give and
teach me.

Thanks be to Christ.

The Prefatory Epistle of Bernard

**Earl of Tresne,
to the noble Doctor and most learned Philosopher
Thomas of Bononia.**

My Friend,

If I had anything more noble, imagine you with what
good will I should dedicate it to thee, for having
considered the wonderful virtue of this Science in
its height, which you are not ignorant of, therefore
was I willing to dedicate this my Labour unto thee,
intreating thee to accept it with as good a will as
I give it unto thee, and conclude that whilst I give
thee this my Labour, that I have given a greater
Treasure than was ever ordained by the good pleasure
of the omnipotent God, according to the course of
Nature.

There is a way truly of arriving to an Universal
Knowledge, which we commonly call the Philosophers
Stone, and thou shalt find it in this my little Book[54]
(little, I say, in words, but great and high in
substance) also it containeth entirely every
Science, that is to say, the beginning and ending.
Thou shalt find this my Book divided into four
parts, and thou mayst judge thereof after thou hast
well understood it. Farewell.

From Tresne, May 12. 1453.
FINIS.

---

[54] Bern. Trevisanus de Transmutatione Metallorum, 4 libris, &
inpressus est cum Jo. Fr. Pici opere de Auro, Ursellis 1598.
80.

The following Epistle I have caused to be printed,
not for the signification there-of either as to
quality or quantity, but only to prevent the
mistaking the one Epistle for the other; and could I
have found more Epistles between these two most
excellent Authors, I should not have hesitated their
publication, but conclude that they would have been
as welcom to our English Philosophers, as any either
Ancient or Modern Writers. Vale.   W.C. Bibl.

A brief Rehearsal of the

Preparation of the Philosophers Stone.

Recipe, and sublime him from his Earthly substance, and then dissolve him into his former substance: then if it be to the Red Work take Sol, if it be to the White Work take Luna, and dissolve it in the said Mercury, until they be both one Mercury, which will not be without Putrefaction; then separate the Elements, and decoct them according to their due proportion. Note, this Sulphur Philosophorum is the Earth of the Elements calcined, sublimed and fixed; then it is coloured with either Sol or Luna, according as thy Work is, the which Sol or Luna is added to fresh or other Mercury after the order of Amalgama; then fixing the Sulphur and the Elements, and that new Sol which is called the Earth, according to their due proportion; the which Names of weight shall not be made mention of here, for the love of him that taught it me, and lest too common it should be; for if it would be named in two Books, then all the World would decay in Husbandry and Industry, if not in Honesty, which I pray God prevent. Amen.

FINIS.

A Word from the Publisher

Thank you for purchasing this small work from The R.A.M.S. Library of Alchemy. During his lifetime, Hans Nintzel was dedicated to the identification, acquisition, study, retyping and, when necessary, translation of what he considered to be the most important known works on Alchemy. Hans was assisted by his sparse network of fellow Alchemists, all members of the Restorers of Alchemical Manuscripts Society (R.A.M.S.). I was an active member of R.A.M.S.

My goal is to publish all of the works originally made available through R.A.M.S. as photocopies. To facilitate this, I have chosen to have the books professionally printed. I also have a few titles that I intend to add to the original R.A.M.S. Library, selected by strict criteria established by Hans.

The works from the original R.A.M.S. Library are republished by R.A.M.S. Publishing Company in the collection, "The R.A.M.S. Library of Alchemy," with permission of the Estate of Hans W. Nintzel.

If you have a work on Alchemy that you believe should be a part of the R.A.M.S. Library, please contact me through R.A.M.S. Publishing Company.

Philip N. Wheeler